REVOLUTIONARY
ENGINEERS

REVOLUTIONARY ENGINEERS

LEARNING, POLITICS, AND ACTIVISM
AT ARYAMEHR UNIVERSITY OF TECHNOLOGY

SEPEHR VAKIL, MAHDI GANJAVI,
AND MINA KHANLARZADEH

THE MIT PRESS   CAMBRIDGE, MASSACHUSETTS   LONDON, ENGLAND

The MIT Press
Massachusetts Institute of Technology
77 Massachusetts Avenue, Cambridge, MA 02139
mitpress.mit.edu

(cc) BY-NC-ND

The MIT Press would like to thank the anonymous peer reviewers who provided comments on drafts of this book. The generous work of academic experts is essential for establishing the authority and quality of our publications. We acknowledge with gratitude the contributions of these otherwise uncredited readers.

This book was set in Stone Serif and Stone Sans by Westchester Publishing Services. Printed and bound in the United States of America.

Library of Congress Cataloging-in-Publication Data

Names: Vakil, Sepehr, author. | Ganjavī, Mahdī, 1985 or 1986- author. | Khanlarzadeh, Mina, author.
Title: Revolutionary engineers : learning, politics, and activism at Aryamehr University of Technology / Sepehr Vakil, Mahdi Ganjavi, and Mina Khanlarzadeh.
Other titles: Learning, politics, and activism at Aryamehr University of Technology
Description: Cambridge, Massachusetts : The MIT Press, [2025] | Includes bibliographical references and index.
Identifiers: LCCN 2024037530 (print) | LCCN 2024037531 (ebook) | ISBN 9780262552196 (paperback) | ISBN 9780262382830 (pdf) | ISBN 9780262382847 (epub)
Subjects: LCSH: Arya Mehr University of Technology. | Engineering—Study and teaching (Higher)—Iran—Tehran—History—20th century. | Engineering students—Iran—Tehran—Interviews. | Tehran (Iran)—Politics and govenrment—20th century. | Nasr, Seyyed Hossein—Interviews.
Classification: LCC T173.T443 V35 2025 (print) | LCC T173.T443 (ebook) | DDC 620.00955/250904—dc23/eng/20250111
LC record available at https://lccn.loc.gov/2024037530
LC ebook record available at https://lccn.loc.gov/2024037531

10 9 8 7 6 5 4 3 2 1

EU product safety and compliance information contact is: mitp-eu-gpsr@mit.edu

CONTENTS

A NOTE ON TRANSLITERATION

We have used the Library of Congress standardization for transliterating Persian words. Whenever feasible, Persian words present in this work are transliterated into their most commonly used English forms. Quoted works within this book may employ various systems of transliteration, and direct quotes are faithfully represented as they were originally written.

PREFACE

The idea for this book was conceived in a specific moment. On a bright, crisp fall day in 2018, I (Vakil) was enjoying a cup of steaming chai at Reza's restaurant in the Andersonville neighborhood of Chicago while reading an article by the prominent historian Stuart Leslie about the history of Sharif University (Leslie and Kargon 2006). Leslie portrays a dramatic meeting in 1973 between the Shah of Iran and Dr. Seyyed Hossein Nasr, a brilliant academic star who was educated in the West but who had risen to the top of the academic ladder in Tehran's other major university, University of Tehran. The Shah wanted to appoint Nasr to be the next chancellor of the university. Nasr was, by any account, a complex and enigmatic figure. He was the first known Iranian to study at MIT in the 1950s, before going onto Harvard where he earned a PhD in the philosophy of science. He hailed from a powerful and strictly pious Iranian family, steeped in an orthodox Islam antithetical to the revolutionary Islam that powered the 1979 revolution and has come to define the current Islamic republic. At the time, he had already acquired a degree of notoriety for his scathing philosophical treatises on Western science and for his perspectives on *Iranianizing* science and technology. During their meeting, the Shah invited Nasr to implement his bold ideas as chancellor of the new university. Nasr accepted the challenge. Nasr has since gone on to become a renowned scholar of Islamic philosophy and currently is a university professor at the George Washington University (GWU). Having learned that he is alive and working in the US, I immediately decided that I needed to meet Nasr in person. Below is the text of my initial email to his assistant, introducing myself and declaring my interest in meeting Nasr:

Dear Amy,

My name is Sepehr—I am an Iranian Assistant Professor of Education at Northwestern, with a B.S and M.S in Electrical Engineering and PhD in Education. I am planning a project examining the experiences of Iranian engineering students at Sharif and University of Tehran. My project also includes an examination of how engineering education developed in Iran, which brought me to the role of Dr. Nasr as Chancellor of Arya Mehr University. I am very excited to meet Dr. Nasr, to learn more about his vision for an "indigenous approach to technology" in Iran, and to learn how engineering education evolved over the last several decades. I would be genuinely honored to speak with him. (email November 5, 2019)

After several such emails and calls to his office, I eventually was contacted by his secretary, who told me that Nasr was aware of my inquiry and had agreed to meet. After coordinating with my wife, I immediately booked my trip to Washington, DC, and met him for our first in-person interview on November 7, 2019. His office was in a large suite located on the seventh floor of the Gelman Library on the GWU campus. I waited patiently as he finished up a conversation in his office, perusing the assortment of Iranian and Islamic art decorating his suite. Several others were also waiting, including a woman who hurriedly asked Nasr for a selfie as soon as he opened his office door. I soon learned that people travel from all over the world to visit the professor. When I was invited to enter his office, I asked him if he preferred to converse in Persian or English, and he confidently stated the choice was mine. Due to the limitations in my own command of the Persian language, I conducted the first interview in English. Our discussion revolved around his recollections as chancellor, including the attempt on his life by militant groups, his philosophical views on science and technology, his thoughts about the spiritual decay of the West, and the fate of Sharif University after the 1979 revolution.

Despite his advanced age, our conversation was fast paced and highly energetic. He was grandiose, at times pompous, yet always sharp and precise with his language and ideas. I can only imagine his presence in the years before the revolution when he was at the helm of what was arguably the most elite university in the Middle East. It felt like I was in the presence of a living legend, whose story held the key toward a breakthrough in unlocking a profoundly radical experiment in engineering education. Below you can see the two of us posing for a picture while in his office.

After our meeting, I sent Nasr the following email expressing my intent to pursue this book project:

Salaam Professor Nasr,

It was an honor to meet you a couple weeks ago in DC. Thank you for sharing more with me about the early days of AryaMehr University. As I mentioned to you, I am an education professor at Northwestern University. I would like to write a scholarly book about engineering education in Iran, and specifically the story of your influence on AryaMehr in the decade leading up to the Islamic Revolution. (November 18, 2019)

I once again flew to DC to meet with Nasr, conducting two more interviews on January 2 and January 3, 2020. As may be expected, Nasr plays a critical role in this book.

RESEARCHER BIASES AND PERSPECTIVES

Before proceeding, as a social scientist, I feel compelled to discuss several biases that informed my initial interest as well as my scholarly engagement in relation to the role of Nasr at Aryamehr University of Technology (AMUT). As a scholar of learning, equity, and culture with a particular interest in computing and engineering education, it was inevitable that I would need to find a way to deeply understand the processes by which an elite engineering university was developed to honor and embrace Iranian culture and Islamic philosophy. My academic training has been anchored in social and cultural perspectives in education. While at Berkeley, my academic advisor was Dr. Na'ilah Nasir, a leading scholar on learning, culture, and identity. Thus, in my first encounter with Nasr, I saw a

0.1   Sepehr Vakil (left), Hossein Nasr (right). Photo credit: Personal collection of Sepehr Vakil.

tremendously exciting opportunity to contribute to this broader conversation about education and culture, specifically from a Middle Eastern/North African (MENA) perspective—and more precisely from an *Iranian* perspective. At the same time, my own spiritual identity exists somewhere between atheist and agnostic. I grew up in a secular Iranian family that was, at least to a degree, skeptical—and even distrusting—of Islamic ideology, specifically of its instrumentalization in Iranian society and politics. So, while as a scholar I was and remain deeply interested in questions of culture and education, in particular, to non-Western approaches to science and technology, I embarked on this project highly skeptical of a supposed "marriage" between Islamic philosophy and engineering education. On the other hand, Nasr's critical views on the ethics of Western science and technology resonate deeply for me and have been important themes that have animated my own scholarship. In a way, Nasr was the opening muse for this project, embodying the conceptual convergences and entanglements, tensions and contradictions that came to characterize not only Nasr but the university as a whole.

I had another set of experiences that informs this project, which trace back to my family's immigration from Iran to the US, as well as the period when I was an engineering student at UCLA, where Sharif students abounded. I met many of these students at UCLA when I was working toward a master's degree, meeting dozens of Sharif alumni who were doctoral students or postdocs. I had entered UCLA myself as a freshman in 2001 just weeks after the horrific events of 9/11.

Not unlike other major universities, UCLA had contributed greatly to the military-industrial-academic complex and had benefited greatly from its relationship with it. Hallways in the legendary Boelter Hall, where I spent many a sleepless night, were adorned with research posters bearing logos acknowledging funding from the Department of Defense, Raytheon, and other weapons manufacturers. The university was regularly tapped for expertise to aid in "counterterrorism," which in its most generous reading is an implicit endorsement of the politics of militarism, war, and empire. In 2003, the United States unlawfully invaded Iraq under the false accusation that Iraq possessed weapons of mass destruction. There is an irony here in how my life and professional trajectories crisscrossed—the US supported Saddam Hussein's Iraq against Iran, thereby forcing my family to immigrate to the West. Less than two decades later, I was studying engineering at UCLA, and the US was once again engaged in war in the Middle East, only this time, the adversary was Iraq. However, one thing remained constant: Both wars were profitable for the US economy and a boon for engineering departments throughout the country.

In the years after the invasion of Iraq, which coincided with the second half of my undergraduate career, antiwar groups proliferated across the UCLA campus, which attracted thousands of students and other Los Angeles residents who gathered on Ackerman Plaza for protests and marches through the campus. I would attend some of these events, mostly on the periphery, my own political identity developing slowly and steadily like a small flame adjacent to the large fire of antiwar radical activism and protest that was rapidly swelling around me. I note here—and suspect this will not be a surprise to readers—that the engineering students and faculty were by and large *completely* detached from these activities. The irony is that the immoral and illegal war against Iraq, which was animating activists at UCLA and throughout the nation, was a great boon for engineering departments—including the Henry Samueli School of Engineering at UCLA. Funding from DARPA (an agency within the Department of Defense with the explicit mission of developing technologies that advance the goals of the US military) and military contractors is an unexamined norm in engineering, computer science, and other STEM disciplines. These relationships, in addition to the racism and sexism that have long characterized the field, add up to an academic discipline that is morally untenable to the minds of many progressively inclined university students. War is big business, especially for research universities, especially for technical and scientific research that sometimes has implicit—and often *very explicit*—implications for technologies of war, surveillance, and control.

After I graduated with a BS in 2006, I entered the graduate program at UCLA under the supervision of Professor Ali Sayed, a man of Arab-Brazilian heritage who was a leading international expert in signal processing, and who is currently the dean of engineering at EPFL in Switzerland. Dr. Sayed's Adaptive Systems Laboratory was home to two Iranian postdoctoral scholars, both of whom had completed degree programs at Sharif University. Our lab was not the exception: Dozens of Sharif PhD students and postdocs worked in various labs and departments in the UCLA engineering school. It is noteworthy not just that so many Iranian students studied in these engineering and technology departments but also that so many of them hailed from one particular institution in the capital of Iran.

I came to know and befriend many *Sharifees* (as they were affectionately called) during my time at UCLA. Back then, I had an overall impression of them as intellectually talented and, considering the aesthetic standards for engineering students, delightfully fashionable. In many ways, I considered them as exceptional, though I did not view them as particularly politicized or politically engaged. In this sense, they were no different from the typical members of UCLA's engineering community. At the time, I did not make anything of this. After all, they were *engineering students*. However, notably, in sharp contrast, historical accounts of Sharif University, as well as public discourse and news coverage about Iranian engineering students, is decidedly political. As a researcher of identity and STEM education in the US, I knew this contradicted the conventional social construction of engineering identities in the West. The unique combination of my Iranian background and academic training, both in engineering as well as in the field of education and learning sciences, had brought into sharp relief a startling contradiction that provided the initial conceptual frame for the investigation we describe in this book.

During our research, we have come to understand that to view the dual identities of engineer and activist as oxymoronic or contradictory, rather than as something quotidian and mundane, is to take on a particularly Western or Orientalist perspective (Said 1979). By this, we mean that to view what is ordinary or normalized in the Iranian context as exotic or peculiar from a Western point of view necessarily positions the Global South as other, as an object of Western fascination. Having said that, it is decisively the case that my curiosity about radicalized engineering students was a key starting point for this project. As a former engineer and engineering student who experienced firsthand the startling political apathy of engineering university cultures in the US, this book represents an effort to understand how in Iran the fundamental paradox between an engineering and a social justice identity is, in fact, not a paradox at all. It is noteworthy that in describing this project, and this departure point for the project (the "paradox" of politicized engineering students), American colleagues quite readily understand, tacitly approving and joining in our sense of wonder at how engineers came to occupy identities as activists. When giving presentations at conferences or lectures at universities, I have come to anticipate reactions of awe and wonder at the mere mention of politicized engineering students. The conceptual shared assumption here, which generates so much curiosity in the first place, is a commonly shared perception of the typical apolitical engineer and engineering student in the West.

ACKNOWLEDGMENTS

We extend our heartfelt gratitude to the countless individuals who have supported the ideas and writing of this book. While many have contributed to this journey, we wish to acknowledge a few groups and individuals to whom we are particularly indebted.

First, we express our sincere thanks to the Spencer Foundation for their generous grant and for recognizing that a historical study of a singular university in Iran has unique relevance for the education field's ongoing efforts to understand the complex interplay among culture, politics, identity, and learning. Funding from the Spencer Foundation was instrumental in supporting our postdoctoral scholars and our extensive archival research, which included research conducted in Iran. The grant also enabled us to convene an exceptional advisory board of widely recognized leaders in their respective fields: Professors Stuart Leslie, Elizabeth Todd-Breland, Kris Gutierrez, and Narges Bajoghli.

We are grateful to the School of Education and Social Policy at Northwestern University (SESP) for providing a highly supportive intellectual home for our research. We also appreciate the support of other university groups, including the Center for Middle East and North African Studies (MENA), the Colloquium for Global Iran Studies (CoGIS) group, and the Technology, Race, Ethics and Education (TREE) Lab.

Without a doubt, this book would not have been possible without the unwavering support of our family members. A heartfelt thank you to kihana miraya ross, Roozbeh Vakil, Sasan Vakil, Maryam Hazeghazam, Mehdi Hazeghazam, Mehrat (Hadi) Hazeghazam, Pedram Mahinpour, and Mehrnaz Mansouri. Thank you for nurturing this work with your steadfast love and care and for your patience as we likely talked your ears off about this project over the last few years! Our profound thanks also go to our generous colleagues who energized and inspired this book in countless ways. Thank you to Amy Slaton, Paymun Zargar, Shirin Vossoughi, Nichole Pinkard, Ali Gheissari, Shahrzad Mojab, Bryan Brayboy, Megan Bang, Hamed Koupai, Reed Stevens, Maryam Athari, Emrah Yildiz, and Hamed Yousefi. We are especially grateful to Elham Beheshti, whose collaboration with Sepehr on a related project focused on the University of Tehran was a vital precursor to this work. We also owe a special debt of gratitude to Na'ilah Nasir, whose visionary scholarship on culture, identity, and learning provided a rich conceptual foundation for our project.

We are deeply grateful to our interviewees, whose profound knowledge and thoughtful perspectives significantly enriched the depth and quality of this work. In addition to the anonymous interviewees, we extend our gratitude to Seyyed Hossein Nasr, Hassan Ghazimoradi, Hossein Amanat, and Shiva Farahmand Rad for their invaluable insights. Special thanks to Firooz Partovi and Farideh Partovi for sharing precious photos from their personal collection, which are included in this book. We also acknowledge Selda Shamloo for her meticulous translation and transcription of the interviews.

Our deepest thanks go to the dedicated staff and archivists at the Massachusetts Institute of Technology Archives, Sharif University's Ganjīnah Historical Document Center, Iran's National Library and Archives, and Iran's Parliamentary Library. Their invaluable assistance and expertise have been instrumental in providing access to critical historical documents and resources that have significantly enriched our research. We also extend our gratitude to the anonymous reviewers whose valuable insights and feedback helped sharpen our arguments and refine the structure of this book. Finally, we are profoundly grateful to the editorial team at the MIT Press, particularly Susan Buckley, for their unwavering support and guidance.

INTRODUCTION

In 1966, the Shah of Iran established Aryamehr University of Technology (AMUT), now known as Sharif University of Technology, as part of a larger campaign to modernize the nation (Leslie and Kargon 2006). Thirteen years later, in 1979, engineering students at AMUT played a critical role in the 1979 revolution that overthrew the Shah and his regime. Sharif University continues to be widely recognized, both for its reputation as the most prestigious engineering institution in the Middle East and for its legacy of radical student activism and protest. Indeed, Iranian engineering students at AMUT have historically engaged in social justice and political activism simultaneously and alongside deep engagement with scientific and technological education (Mojab 1991). Notably, Sharif students have been at the epicenter of recent protests in response to the state murder of a twenty-two-year-old Kurdish-Iranian woman, Mahsa (Zhina) Amini. Unlike perceptions of engineering in Western societies, engineering students in Iran are widely understood to be the most politically engaged students. In fact, it is not considered unusual to be an engineering student and to also be an activist, intellectual, artist, or politician.

This is in stark contrast to how the identities of engineering students and professionals are typically viewed in the US. While many changes have occurred over time with regard to how the average American engineer views ethics and matters of social responsibility (Layton 1986) and how the

American public views the engineer, the stereotypes attached to the engineering profession and engineering students have been relatively stable in America since the early twentieth century. However, the apparent stability of what engineering represents in the public mind belies a more complicated and interesting history that is relatively unknown. In the 1960s and 1970s, leftist engineers founded an organization named the Committee for Social Responsibility in Engineering, and many other engineers were directly involved in the antiwar and civil rights movements of that era (Wisnioski 2012). In the archival research we conducted for this book, we have encountered striking examples of activism by Massachusetts Institute of Technology (MIT) engineering students who protested against the tacit support and involvement of their institution with the Shah of Iran in his quest for nuclear weapons (*The Tech*, 29 April, 1975), as well as against other perceived ethical transgressions.

Earlier in the twentieth century, many progressive American engineers had struggled against conservative elements to reform the profession. Such activities have been well documented. A prominent example is the book by historian Ronald Kline about electrical engineer and socialist Charles Steinmetz, who later became president of General Electric (Kline 1992). Ultimately, these efforts were generally unsuccessful (Noble 1977) in shifting either the profession or its image in the eyes of society. As a result, the overwhelming scholarly consensus, as well as the attitudes of the public, is largely unified in viewing engineering as a profession that is practically inseparable from corporate America and the military-industrial-academic complex (Giroux 2015). To the extent that engineers in the United States are connected to these structures of power and influence by virtue of their professional associations, their individual politics have become interwoven with the politics of American capitalism and militarism. A very interesting and important line of academic work has shown engineers in the US lean conservative compared to their colleagues in the social sciences and humanities (e.g., Moore 2020). Sharif University, formerly AMUT, given its dual history of radical politics and elite technical talent, shatters these settled conceptual categorizations of engineering identity common across Western societies. At a fundamental level, this book closely examines the formation of an elite engineering university in revolutionary Iran and the individuals who shaped and were shaped by it.

While Sharif is widely known and respected in Iran and by the Iranian diaspora, non-Iranians (particularly Americans) may not be familiar with Sharif University. Many graduate students and current faculty at top-tier institutions in the US—especially those in the sciences and engineering fields—are Sharif alumni. Dubbed by many as "the MIT of the Middle East," Sharif has provided tremendous benefits to the US through the brilliance and ingenuity of these Iranian immigrants. For example, a Sharif alumnus named Maryam Mirzakhani, the late Iranian mathematician and Stanford professor, was honored in 2014 with the most prestigious award in mathematics, the Fields Medal—the first Iranian (and only the second woman) in history to win. Another notable Sharif alumni is the widely known engineer and the former chancellor of Sharif, Mehdi Zarghamee. After 9/11, Zarghamee was the principal investigator on a National Institutes of Science and Technology (NIST) project that employed structural modeling techniques to analyze the collapse of the World Trade Center. It is remarkable that while many Iranian students have been denied student visas based on immigrant bans, one of the most prominent Iranian-born engineers of the twentieth century was leading the highest-profile government-funded study to scientifically assess the damage in the aftermath of 9/11. It is difficult to conceive of a more patriotic act. Likewise, we might consider another Sharif alumni, Firooz Partovi, who was the founding professor of the physics department at AMUT. His twin sons, Hadi and Ali Partovi, are the co-founders of Code.org, an organization that has elevated computer science education to a top national priority for K–12 public schools in the United States. While Hadi and Ali are not themselves Sharif alumni, the powerful influence of their father as a widely respected AMUT physics professor on them is unmistakable. In essence, two of the leading figures in American computer science education are Iranian-born immigrants directly linked to the legacy of AMUT.

These notable Sharif alumni make evident that, at least in part, the story of Sharif is connected more broadly to the internationalization of higher education. Engineering graduate schools in Europe and the United States are brimming with immigrant students, particularly from Global South countries such as Iran, India, China, Brazil, Argentina, and Nigeria. Consider these staggering statistics from a recent 2022 report by the National Foundation for American Policy:

In U.S. graduate programs in science, technology, engineering and mathematics fields [the research] finds that foreign nationals account for 82 percent of full-time graduate students in petroleum engineering, 74 percent in electrical engineering, 72 percent in computer and information sciences, 71 percent in industrial and manufacturing engineering, 70 percent in statistics, 67 percent in economics, 61 percent in civil engineering. (Redden 2021)

It may be tempting to compare AMUT to India's elite IIT system (Subramanian 2019), another significant global institution that nurtures hundreds (if not thousands) of scientists and engineers that feed into the institutions of Western government-industry-academia, which relies on foreign technoscientific talent. The policy report cited earlier describes a trend, predating the COVID-19 pandemic, of declining international enrollments in US institutions, particularly relative to those in Canada. Reflecting on this decline, the report states:

That is a worrisome sign for America's technology future due to the vital role Indian immigrants play in science and engineering in the United States. U.S. restrictions on international graduate students from China are another problematic policy development likely to harm U.S. companies, universities and the American economy. (National Foundation for American Policy, March 3, 2022)[1]

Indeed, the politics of higher education have served as a critical site for US global strategy and talent acquisition. Tapping talent from poorer, and often browner, countries has been part of the US playbook for decades. Yet, the case of Sharif University does not neatly fit into this narrative. The early history of Sharif reveals a crucial difference from institutions such as the IIT system of India. Beyond the direct involvement of many students and faculty in the popular revolt against the Shah, the early architects of Sharif University were motivated by a vision to establish a modern engineering university that was rooted in Iranian cultural values and would be designed specifically for Iran's benefit. Even the Shah himself, despite his friendly relations with the West, envisioned an engineering institute developed precisely with concern first and foremost for the future of Iran. However, for the Shah, collaboration with Western universities was not at all contradictory to this Iran-first agenda.

This richly complex historical context provides a unique opportunity to contribute to several intersecting lines of research in the social sciences and humanities. In the field of education, and more specifically within the learning sciences, STEM education, and science and technology studies, a vibrant body of work has challenged the cultural biases inherent in these disciplines. A broad recognition has arisen of the urgent need to reimagine STEM education in ways that honor the traditions, beliefs, practices, and epistemologies of nondominant groups. For example, ethnomathematics is a field dedicated to understanding the mathematical practices of diverse cultural communities, often with an emphasis on those that have been most marginalized and excluded from mathematical opportunities (d Ambrosio 1985). In science education, scholars have shown that deep engagement with Indigenous practices and ways of knowing is not just ethical but a global priority, considering the scale and scope of ecological crises the world is facing (Bang, 2020). In computing and engineering education, similar initiatives aim to honor and leverage Black, Indigenous, and Latinx values and perspectives toward equitable and culturally sustaining forms of learning (Ryoo et al. 2020; Rankin and Thomas 2020).

Adjacent to these, a considerable body of work has examined the role of values in the formation of STEM identity for underrepresented students. It is well established, for example, that perceptions of science as White and masculine play a critical role in the reproduction of race and gender inequity (Blickenstaff 2005; Brickhouse et al. 2000; Carlone and Johnson 2007; Margolis 2013). Supporting a multiplicity of epistemological values and recognizing the historically accumulated "settled" values present in dominant representations of science has been a critical area of more recent work in science and environmental education (Bang and Medin 2010).

In technology fields, a pervasive culture of competition and individualism has been shown to discourage students who have more altruistic or social justice ambitions (Carlone and Johnson 2007; Garibay 2015; Pawley 2009). In an ethnographic study, Stevens et al. (2008) present in-depth case studies of two women of color who both, though being declared engineering majors, viewed engineering as lacking relevance for their personal development. In a similar vein, recent work has shown that negotiating one's current and future identity in STEM is mediated by "perceptions about self, science, and scientists' work" (Kang et al. 2019, 421).

All of these examples are part of an even broader recognition that equity and social justice are deeply tied to questions of culture, power, and knowledge. Calls to decolonize the curriculum demonstrate that equity in education can no longer be viewed merely as matters of inclusion, diversity, or access. One must ask, "Access to what?" The pioneering work of scholars such as Gloria Ladson-Billings and Carol Lee have advanced the idea that "good teaching" for Black students in the US entails nothing less than pedagogies that are carefully attuned to the cultural and linguistic practices of African American children, along with curriculum that is both relevant and rigorous.

RECASTING QUESTIONS OF DIVERSITY, EQUITY, AND JUSTICE IN STEM/ENGINEERING EDUCATION

Unsurprisingly, engineering-related industries ranging from Silicon Valley's Big Tech companies to the giants of American militarism such as Raytheon, Northrup Grumman, and Boeing have deeply entrenched racial and gender inequities. It has become almost axiomatic to observe that these industries are largely composed of White men, an inevitable result of a profession molded in the cauldron of American racism and heteronormative patriarchy. Ironically, the diversity that does exist within the engineering profession is largely due to socioeconomically privileged African, Asian, South Asian, and Middle Eastern immigrants, including many hailing from Sharif University. The growing presence of these immigrant groups is often packaged and sold as evidence of progress toward inclusion, concealing the enduring exclusion of poor and working-class Black, Latinx, and Indigenous communities. Yet, despite these unassailable demographic realities, diversity, equity, and inclusion (DEI) efforts in tech and engineering industries are commonly framed by conservatives as evidence of "woke politics"—an alleged progressive conspiracy to undermine the freedom of (White) Americans. For instance, referencing recent problems on Boeing airplanes, Elon Musk made several baseless statements accusing diversity efforts of making flying less safe. Quite to the contrary, scholars have shown that DEI initiatives have been largely "performative"—superficial and ineffectual endeavors unable to substantively redress legacies of discrimination against individuals who are Black, Indigenous, or People of Color (BIPOC) (Ahmed 2020; Slaton 2010). Though significant, these findings are also unsurprising, as many efforts to promote DEI in engineering (and STEM education more broadly) have neglected to examine how the politics and ethics of the engineering profession itself may be a major "deal breaker" for students with marginalized identities who often demonstrate aspirations for social justice (Garibay 2015; Vakil 2020). This is despite the fact that decades of research in education, psychology, and learning sciences have demonstrated that student interest in and identification with STEM domains is dynamically linked to how they encounter the cultural norms, practices, and narratives of the disciplines.

In other words, researchers have found some students may choose to reject engineering or other STEM pathways, not due to lack of interest or adequate academic preparation but rather because of a perception that involvement in these fields would be a form of "selling out." This kind of analysis recasts the problem of equity and diversity in STEM from one that has long centered on the deficits of students from marginalized communities to the deficits of the disciplines (and the career pathways they are linked to) themselves. Fortunately, we are beginning to see a growing awareness of and engagement with these issues in the K–16 STEM educational community in the US, with increasing emphasis on ethics, social responsibility, and civic engagement. In fact, given the political, social, and ecological implications of science and technology across global contexts, there is an urgent need to understand how engineers and scientists become sensitized to ethical considerations, and how best to address the stakes, consequences, and moral dimensions of science and technology in modern STEM education.

THEORETICAL FRAMES

We ground ourselves within sociocultural and cultural-historical methodological perspectives (Cole 1996; Lee 2008), as well as interpretive and ethnographic traditions in education (Erickson 1986; Packer 2010) attuned to the complex interplay of history, power, discourse, and learning. Drawing on diverse theoretical traditions that span psychology, sociology, and anthropology, sociocultural perspectives in the learning sciences have centered the role of identity in the learning process (Lave 1993; Mehan et al. 1996; Nasir and Hand 2006; Wortham 2006). At the heart of identity research lies a recognition that supporting learning requires empirically and theoretically rich accounts of who students are as learners and as people, and who they might become through their future experiences in schools and learning contexts more generally. The notion that learning environments can either support or constrain particular identities, or "imagined trajectories of becoming" (Nasir and Hand 2006, 468) is anchored in theorizations of learning as a fundamentally cultural process (Bang and Medin, 2010; Cole 1996; Nasir et al., 2020; Rogoff 2003). Seeking to build on this work while simultaneously exploring uncharted theoretical terrain in the learning sciences, there have been recent calls to more explicitly conceptualize and empirically study the impact of power, politics, ideology, and ethics on learning and schooling (Esmonde and Booker 2017; Mckinney de Royston and Sengupta-Irving 2019; Philip et al. 2018; Vossoughi et al. 2020).

With these considerations in mind, one of the principal objectives of this book is to explore the ways in which the political identities of engineering students at AMUT were dynamically linked to the specific national, political, and cultural context of Iran in the years immediately preceding the 1979 revolution. This is also in line with recent calls for research into activist cultures among engineers and engineering students, with keen attention to diverse global contexts in specific historical moments (Martin et al. 2021; Downey and Lucena 2005).[2]

METHODOLOGICAL NOTES

Our project delves into the transient history, cultural, and political atmosphere that profoundly influenced the life and identity of those associated with the AMUT campus from the time of its founding until the 1979 revolution. The political history of AMUT and its activist movements is characterized by a culture of secrecy among the activists, countered by brutal suppression and paranoia from the state and its intelligence services. The challenge lies in reconstructing a history in which understanding political affiliations often occurred posthumously or after individuals were incarcerated. In addition, as we discuss later in the book, organizational instructions from militant groups strictly mandated that student cells remained isolated from each other, forbidding any overt friendly relations on campus. It is essential to recognize that the history and memory of their struggle have been suppressed by the Pahlavi regime and selectively remembered and memorialized by the postrevolutionary Islamic State. Unraveling this politically fragmented, suppressed, and distorted history demands a meticulous investigative process involving data gathering, cross-checking, and critical reevaluation of narratives and archival records.

In exploring this complex history, we rely on archives and oral histories and consider their complementary roles and contrasting perspectives. On one hand, we are facing damaged and incomplete archives. The AMUT institutional archive, according to its brief finding aids,[3] has either not collected all the instances of political documents, or if it has, is not willing to share them with researchers.[4] Thus, while the institution possesses several textbooks and educational and university publications, it has not been collecting documents on the political life of the university. Fortunately, many historical documents, including the reports by SAVAK (Sāzmān-i 'iṭilā'āt va 'amnīyat-i Kishvar, Bureau for Intelligence and Security of the State) and those by certain faculty and chancellors to the office of prime minister, have been preserved in Iran's National Library and Archives. Iran's Parliamentary Library also preserves several student bulletins and journals from this period. The Center for Historical Documents Survey, a postrevolution, state-affiliated organization, has published volumes of documents from the archives confiscated from SAVAK. These documents include many in which SAVAK was reporting on the activism at AMUT. The Muslim Student Followers of the Imam's Line has also published the US embassy's confiscated records in eleven volumes, translated into Persian, some with a copy of the original document. The US Department of State has also declassified several such documents in recent years. These diverse archival sources, along with historical dailies and journals, help with putting the missing pieces together.

Even so, when it comes to the documents of Marxist political organizations, the archival materials on the subject are not only scattered but also have been suppressed or removed from national institutional archives. Thankfully, two political groups, the Organization of Iranian People's Fedai Guerrillas and the Tudeh Party, have digitally preserved and provided access to their key documents and pamphlets on their respective websites. The Archives of Iran Opposition Organizations, which is an online collection, has also made accessible many related documents, which we have utilized in our investigation. Notwithstanding all of these, we are still aware that activism under political suppression generally necessitate the destruction of any documents that could have put other members of a movement in danger. Although necessary, such practices have a devastating effect on later efforts by researchers to reconstruct the history of such movements based solely upon examination of archived material, so we have had to use other approaches as well, such as examining the published memoirs and interviews with activists, students, faculty, and staff to locate data on unarchived events and developments.

Also, over a hundred student activists at AMUT (the total estimated as 135 by Farahmand Rad) were killed during the revolutionary struggle, but this number is probably still far from certain. Because many of these people did not leave documents or written memoirs about their experiences, their voices have been lost. Certain institutions, including Sharif University, have also compiled oral histories of interviews with alumni and former chancellors. While these oral histories contribute to a more detailed understanding of life within specific institutions, it is essential to approach them with caution, for the voices represented within them are confined to those conforming to the general boundaries of the official historiography of the Iranian revolution. For instance, none of the oral histories conducted by Sharif University include the voices of individuals who were purged or forced to live outside the country in the aftermath of the Islamic cultural revolution. Furthermore, no effort has been made to include the perspectives of leftist alumni in these histories. Our research both benefits from the aforementioned recent developments in data usage and extends data collection in two ways: first, through attempts to locate and discuss materials housed in several archival institutions inside and outside Iran, and second, through the original collection of oral histories intended to cover various aspects of the political, cultural, and pedagogical experiences of learning within AMUT. In addition, in terms of the subject of inquiry, this investigation focuses on several key questions that have rarely been analyzed in the field of educational research on Iran. Notably, it addresses the issue of politicization as a learning process, as well as the relationship between student activism and political opposition groups, as experienced by students and faculty members of the time.

To gain a more nuanced picture in our oral histories, we have interviewed former students and faculty, senior leadership of relevant
institutions—such as Dr. Hossein Nasr and Dr. Firooz Partovi (former head
of the Physics Department at AMUT)—and the architect who designed
the first buildings at the AMUT, Hossein Amanat. We conducted a total of
seventeen oral histories designed to understand the experience of being
a student, faculty member, and administrator at AMUT during a specific
sociopolitical-historical period. These interviews include nine interviews
with alumni, four of which were with women; six interviews with former faculty members, including one with Firooz Partovi, the first head
of Physics Department at AMUT, and one with a female faculty member;
three interviews with Seyyed Hossein Nasr; and one interview with Hossein Amanat. Names of all interviewees have been anonymized, with the
exception of five: Hossein Amanat, Seyyed Hossein Nasr, Firooz Partovi,
Shiva Farahmand Rad (alumnus), and Hassan Ghazimoradi (alumnus).
Each provided their consent for us to use their real names. Typically, these
interviews were conducted in Persian and averaged 2.5 hours in duration,
with some exceptions. For example, Hassan Ghazimoradi, whose case we
highlight in chapter 6, was interviewed by us on three separate occasions,
which yielded nearly seven hours of interview data. The interviews were
conducted remotely and were all audio-recorded. All interviews were then
transcribed and translated. Next, we simultaneously listened to the audio
recordings while reading the transcripts and added minor modifications to
the English translation as needed. We independently open-coded all tran-
scriptions, which led to several analytic categories representing our shared
interpretation of the data (Emerson et al. 1995). We conducted careful
analyses of the interviews, entailing multiple rounds of close listening, the-
matic and interpretive coding, memo writing, and collective sensemaking.
A central question we were investigating during the interviews with former
alumni was: "What did it mean to be a politicized engineering student at
AMUT between 1966–1979?" As social scientists, we approached this question carefully and began with the important recognition that our answer
was necessarily informed by insights gained from close listening and analysis of our oral histories. And, as with any sample, our sample contains some
inevitable biases. Although our respondents provided a robust image of

the complex set of social practices, norms, and values that defined the political culture of AMUT, there were limitations. For instance, only one of our interviewees was formally associated with a leftist political group described in depth in chapter 4. This individual was imprisoned for his political activity and never finished his studies. The other AMUT alumni we spoke with were (and, in some cases, still are) sympathetic to leftist movements but were never formal members of the Organization of Iranian People's Fedai Guerrillas (OIPFG) or the People's Mujahedin Organization of Iran (PMOI) during their years at AMUT. They all completed their degrees and went on to have successful careers, some in Iran and many others in the West. Furthermore, even though they did not hold formal memberships or leadership positions, they expressed sympathies with the revolutionary student culture, which was dominant among student activists on the campus, as we describe in chapters 4 and 5. In addition, most of them mentioned participating in boycotts or protests at different points during their college years. And notably, during their college years, none of our interviewees aligned themselves with the Shah or with Islamic groups, even though certainly some students with pro-Shah or Islamic sympathies did exist during this era at AMUT. For many reasons, it is difficult to know precisely the numbers or proportions of students who had formal membership in any political organization. Based on our in-depth analysis of archival records and oral histories, we contend that our sample provides insights that reflect a recognizable strand of experience for students at AMUT during that historical era.[5]

While oral histories offer valuable insights, we further acknowledge the complexities of memory, bias, and trauma. The interviews were conducted between 2019–2024, approximately forty to fifty years after the period under investigation. The lives of some of the interlocutors changed dramatically after the revolution—in some cases, they never returned to the country after the revolution. Furthermore, not all interlocutors had vivid memories of the events or even clear memories of the chronology of events. Moreover, the political and ideological orientations of most of the interviewees have changed considerably during the past several decades.

As diasporic Iranians, our research is at once personal and political—the intellectual soil of our analysis rests upon our relation to and deep connectedness with the themes and subjects of our collective inquiry. In particular, we bring to this work a desire to better understand the experiences of our families leading up to the 1979 revolution and its aftermath and to ultimately deepen our own understanding of historical processes that directly and profoundly shaped our lives and career trajectories. Even so, our analysis also bears the mark of our perspectives as scholars of Iranian descent living in the US and Canada. Our "Iranian-ness" is deeply present in form and function—in our extensive use of narrative and storytelling, and in our collective emotional and spiritual connection to the people, dreams, lands, and waters of Iran.

OVERVIEW OF CHAPTERS

In chapter 1, we begin by describing the pedagogical and institutional history of the university, emphasizing the conditions of its formation and how the new university was central to the Shah's modernization and nationalization efforts. In chapter 2, we focus on a crucial international partnership that supported the new university: namely, the partnership with MIT. We underscore the AMUT–MIT partnership as a hidden history of US–Iran relations but also one that was fraught with complexities and contradictions, including protests by MIT faculty and students, who opposed the partnership with the Shah's autocratic regime. In chapter 3, we turn to Dr. Seyyed Hossein Nasr, the enigmatic Western-educated Islamic philosopher and fourth chancellor of AMUT, who aspired to Iranianize the university and who faced severe opposition by students, including an attempt on his life. We carefully assess his philosophical positions on science and technology and how his searing critiques of Western paradigms shaped his approach to leadership at AMUT. Next, in chapter 4, we contextualize student and faculty activism at AMUT in a longer tradition of university protest in Iran. In the case of AMUT, of course, the protests rapidly swelled, coinciding with the massive popular protests that culminated in the 1979 revolution that overthrew the Shah's regime. Our analysis of activism at AMUT includes a close examination of key texts and ideologies of the student movement and its complex relations with Marxist and Islamist groups, as well as its flaws, missteps, and oversights. In chapter 5, we critically engage the role of gender in the revolutionary culture of AMUT and highlight the experiences of female engineering students. In doing so, we illuminate the dominant masculine culture that prevailed not just at the university but also within the political culture of the time. In chapter 6, the final chapter, we attend closely to the identities and experiences of engineering students at AMUT during the historical period under study. What did it actually mean to be a politicized engineering student at AMUT in the years directly preceding the 1979 revolution? We explore this through an in-depth case study of one of our interviewees, Iranian intellectual and writer Hassan Ghazimoradi. In conclusion, we highlight key implications of this book for several areas of scholarly research, as well as ongoing protest movements in Iran, the United States, and beyond.

1

POLITICAL AND PEDAGOGICAL HISTORY OF ARYAMEHR UNIVERSITY OF TECHNOLOGY (1966–1979)

How is it that the premier engineering university in Iran, and arguably in the broader Middle East region, is simultaneously an institution synonymous with radical student foment for the past sixty years? How did a university that the Shah of Iran viewed as one of his signature projects aimed at modernizing the Iranian nation become a catalyst of the revolution that ousted him from power? To answer these questions, we start with a historical account of the context in which AMUT was established, followed by an institutional and pedagogical history of the university. We illuminate how the formation of AMUT was connected to a set of historical and political conditions both inside and outside of Iran.

AMUT EMERGES IN THE CONTEXT OF THE COLD WAR

The border between Iran and the former Soviet Union stretches across thousands of miles of land and sea (Smolansky 1981). While relatively less recognized Iran was one of the first places where the tensions rapidly intensified between Western and Soviet forces. The allied forces had agreed to leave occupied Iran no more than six months after the end of the war. However, by the end of May 1945, the Red Army had not retracted its forces and instead had provided military support to two newly formed provincial governments in Iran. The Iranian central government perceived this event, commonly referred to as the Azerbaijan crisis (Fawcett 2014), as a direct violation of its national sovereignty.

In response to pressure exerted by the United States, the Soviet Union yielded to international demands and eventually withdrew its troops. Consequently, within a span of a few months, the Iranian armed forces quelled the dissent from provincial governments, thereby reestablishing firm dominion over both territorial and governmental affairs. Of course, efforts by foreign powers to exert an influence on Iran did not begin or end with the Cold War. Throughout the latter half of the nineteenth century, Iran was subjected to the influence of various foreign powers (notably Russia and Britain), which sought to shape the destiny of the nation for their own benefit. Despite avoiding formal colonization, Iran faced two instances of foreign occupation during the First and Second World Wars. Allied forces used an exaggerated account of the Iran–German economic and political relations as a pretext and occupied Tehran in 1941, which resulted in the abdication of power by Reza Shah and the transfer of power to his son, Mohammad Reza Pahlavi (Kashani-Sabet 2023).

Between 1945 and the 1953 coup, the official foreign policy of Iran centered on maintaining a neutral stance toward both capitalist and socialist blocs (during the tenure of Prime Minister Mosaddegh, this approach was commonly referred to as negative equilibrium) (Abrahamian 2001). Meanwhile, the communist Tudeh Party, which had originated in Iran during the occupation period, grew into the most structurally organized political party in the country. Backed by Soviet influence, this party effectively orchestrated the mobilization of numerous student activists, particularly at the University of Tehran (Amirkhosravi 2020). After an assassination attempt on the life of Mohammad Reza Shah Pahlavi at the University of Tehran in 1949, the Tudeh Party was officially prohibited, further straining the already tense relationship between the Shah and the University of Tehran. As we will see, universities would continue presenting very serious challenges to the authority and legitimacy of the Shah, but, nevertheless, they were an indispensable component of his broader ambitions to ramp up the industrialization and modernization of Iran.

In August 1953, nationalist Prime Minister Mosaddegh was overthrown in a coup sponsored by the CIA and MI6. After the coup, the Shah, who had previously fled the country, returned to Iran. After his return, the orientation of the regime shifted toward closer relations with the United States. In addition, with regard to its internal politics, the regime then shifted toward suppressing both nationalist and leftist groups within Iran, further restricting the already limited political liberties in the country. Furthermore, the Shah sought to employ policies to align the educational system of the country with his political and economic plans (Mojab 1991). Previously hindered by the politics of the Mosaddegh era, following the coup, US–Iranian educational cooperation experienced a remarkable resurgence and formed an important element to the US containment strategy during the Cold War era (Shannon 2017; Ganjavi 2023).

THE DEMAND FOR IRANIAN ENGINEERS: POLITICS OF INDUSTRIALIZATION AND IRAN'S WHITE REVOLUTION

The two Shahs of the Pahlavi dynasty, Reza Shah and Mohammad Reza Shah, both believed the progress of Iran depended on major industrial projects, such as the national railroad, steel industry development, and nuclear power initiatives. These undertakings aimed to address the industrial growth needs of the nation. However, despite efforts to attract foreign investment, by the early 1960s, technological transfer remained marginal. The Iranian state faced a significant shortage of engineers necessary to execute its industrial ambitions. By 1965, the country had only approximately 12,000 engineers, amounting to a mere one-sixth of the engineering workforce population in nearby nations such as Turkey (*Rūzigār sharīf* 2016, 14).

The mid-1960s witnessed Iran's initial efforts to industrialize, initially prompted by economic hardships. However, the subsequent surge in international oil prices during the early 1970s provided the Iranian state with a temporary influx of financial resources, enabling it to ambitiously pursue various industrial projects. This era witnessed the establishment of several companies in numerous crucial sectors such as steel, copper, and aluminum.[1]

The Isfahan Steel Company may be the most symbolic example of Iran's moves toward industrialization during this era. According to Mohammad Yeganeh, a former minister of finance in Iran who later assumed the role of the head of the central bank of Iran, the desire to establish a steel company had long been a cherished aspiration for the Iranian people. Although previous attempts had been made, they paled in comparison to the magnitude and scope of the Isfahan project. The establishment of the Isfahan Steel Company not only marked a significant step toward industrialization within Iran but also signified a notable shift in the economic alliances of the nation on the international stage, with Iran turning to the Soviet Union for support (Yeganeh, interview with HOHP).[2]

It is evident then, that the establishment of AMUT, with its charter ratified just five days prior to Iran's parliment's approvement of the Iran–Soviet agreement for the construction of Isfahan Steel Factory on January 13, 1966, was one of Mohammed Reza Shah's signature initiatives to address this pressing need for engineering education and expertise. The connection between the founding of AMUT and the Shah's ambitions for industrialization are abundantly evident.

Yet, there is another aspect to the story. AMUT also was born in the context of what is known as Iran's White Revolution, which encompassed a wide range of reforms initially grounded in six principles, addressing issues such as agrarian reform and women's suffrage.[3] Subsequently, the scope of the reforms expanded to include a total of nineteen principles, comprising administrative and educational transformations. These reforms carried profound economic, social, and political implications, particularly regarding land reform.

In addition to the impetus to industrialize, the establishment and progress of AMUT were intricately intertwined with the transformative impact of the White Revolution in several notable ways. Second, a key aspect of the White Revolution centered on educational reform, prompting the Shah to assume an active role as an educational leader.

From the Shah's perspective, the connection between AMUT and the Isfahan Steel company was so explicit that he even planned to establish the university in Isfahan (*Rūzigār sharīf* 2016, 28). However, Mohammad Ali Mojtahedi, the first chancellor of the AMUT—to whom we would return soon—argued that the university should be located in Tehran, at least temporarily. Over the years, this early tension evolved into a highly politicized issue, leading to significant faculty activism against the government (Zarghamee, interview with HOHP).[4]

According to Zarghamee, the fifth chancellor of AMUT, the establishment of AMUT coincided with significant educational transformations occurring in France and widespread student protests against the Vietnam War in the United States. Majid Rahnama (1924–2015), the minister of science and higher education at the time, urged the Shah to prevent these protests from spreading to institutions of higher education. This led to fundamental changes being introduced in the higher education system (Zarghamee, interview with HOHP).

Thus, as a result of these global student events, as well as the White Revolution, the Shah emerged as a prominent figure in educational leadership, presiding over annual national conferences called the Ramsar Conferences, with the first taking place in the summer of 1968. These annual conferences were centered on a set of twelve articles outlining the guiding principles and objectives of the educational revolution, which were designed to shape and organize the educational development of Iran (Pahlavi 2009 [originally published in 1977], 128). These principles included engaging in dialogue with students, promoting a conducive learning environment within universities, and granting students the freedom to choose from a few electives among their courses. The Shah discussed his educational ideas in his book, *Toward the Great Civilization*:

In Iran's future, those responsible for each job should have both scientific and moral capabilities to fulfill their responsibilities. This means that they should not only be conscious of the secrets of knowledge and techniques but also have integrated their knowledge with vision and great humane values. (Pahlavi 2009 [originally published in 1977], 128)

THE SHAH OF IRAN ESTABLISHES AMUT

On January 8, 1966, the charter of AMUT was officially ratified. Shortly before ratification, the Shah had extended an invitation to Mohammad Al Mojtahedi (1908–1997), then serving as the dean of Alborz College, proposing the establishment of a university focused on science and technology but anchored in the pressing industrial and technical needs of Iran. Mojtahedi had earned his Docteur d'Etate in mathematics from the Sorbonne (Zarghamee 2011) and was a capable and diligent educational administrator who had achieved success in leading the renowned Alborz

College in Tehran, the most famous and influential Iranian college of that time. It was initially an American college established by American Presbyterian missionaries in 1873 but was later nationalized in 1940.[5]

After the completion of the AMUT charter, which designated it as a Royal University falling under the direct trusteeship of the Shah, the chancellorship of the university was bestowed upon the Viceroy trustee (nāyib tawlīyat), a position appointed by the Shah as the highest authority.[6] On November 22, 1965, the Shah appointed Mojtahedi as the first chancellor of the AMUT (*Kārnāmah dah sālah fa'āliyat dānishgāh ṣan'atī* 1977, 13). The Shah made the following formal announcement declaring the establishment of AMUT:

The advancement of Iran's economic development and industrialization endeavors necessitates a skilled workforce aligned with the requirements of a future industrial society. As these programs progress, encompassing the establishment of heavy industries, the demand for qualified professionals to oversee them will surge. To meet this objective, alongside the establishment of Pahlavi University in Shiraz, we have made the decision to establish a well-equipped technical university. This institution will possess the capacity to educate experts with diverse scientific perspectives and technical proficiencies, while also conducting scientific and technical research at a level commensurate with renowned institutions in developed nations.

In selecting professors for this university, we will enlist individuals of great merit who have chosen to settle in Iran, as well as the most accomplished Iranian students who have completed their studies with distinction at prestigious universities abroad. Personally assuming the trusteeship (tawlīyat) of this educational establishment, we have determined that the university shall be named "Aryamehr University of Technology" and will commence its operations on the 1st of Mehr in 1345 [September 23, 1966], following the completion of necessary preparations. (*Nashrīyah dānishgāh ṣan'atī 'āryāmihr bi munāsibat 'avvalīn sālgard ta'sīs dānisgāh*, 11 aban 1345; NLAI-264-027688)

Several months later, on November 2, 1966, during the official inauguration of the university, the Shah, Queen Farah, and other government officials were present. During the visit, the Shah toured various parts of the university and had lunch with the students. Notably, the Shah sat with two students from the university, engaging in conversation with them, both of whom (seemingly) expressed gratitude for the opportunity to study there (Mojtahedi, interview with HOHP). The Shah quoted the students in his speech at AMUT, expressing his happiness and satisfaction

1.1   Shah holding an official meeting with AMUT professors at Gulistan Palace, on October 26, 1966. Photo credit: Nashriyah dānishgāh ṣan'atī āryāmihr bih munāsibat awwalīn sālgard ta'sīs dānishgāh, 1345; NLAI-264-027688.

with the university and its management (*Nashrīyah dānishgāh ṣan'atī āryāmihr, duvvumīn sālgard ta'sīs*, 11 aban 1346, PLI 11–0620—25362).

This event, during which the Shah sat in a friendly manner alongside adoring students at his prestigious new university, is in stark contrast to the negative opinion of him students would later express. The shifting dynamic was clearly visible only six years later, on May 4, 1972, when, as will be discussed in chapter 4, the police were called to the campus to suppress a protest by students and faculty members. Zarghamee, one of the later chancellors of the university, observed that this day marked the first instance that the chant "Down with the Shah" was heard on the AMUT campus and an early time in the political sphere of Iran in general (Zarghamee, interview with HOHP).

1.2 Shah visiting the newly built laboratories at AMUT, October 1966. Photo credit: *Nashriyah dānishgāh ṣan'atī āryāmihr, duvvumīn sālgard ta'sīs,* 1967, Parliamentary Library of Iran (PLI) 11–0620—25362.

MOJTAHEDI: AMUT'S FIRST CHANCELLOR

During its first thirteen years of operation (1966–1979), AMUT underwent significant development and progress. Throughout these formative years, the institution witnessed the appointment of six chancellors, each of whom played pivotal roles in shaping its trajectory. From an educational administrative standpoint, the operational expenses of most universities were largely dependent on financial support from the government. However, government funding for universities encountered bureaucratic obstacles, which impeded faculty recruitment, proper salaries, and research equipment acquisition.[7] In a 1988 interview as part of the Harvard Oral History Project, Mojtahedi recounted the events when the Shah entrusted him with the task of founding this university. With the help of Mohammad Hossein Adib, one of his colleagues at Alborz College, Mojtahedi drafted a charter for this new university. His account reveals that while the initial concept for the university originated from the Shah, the Shah did not envision it as a royal institution at first. Mindful of the financial and administrative limitations that universities under Iran's Ministry of Education were facing, Mojtahedi proposed the university to be independent from the Iranian Ministry of Education but under the patronage of the Shah himself. He even suggested the name to include *Aryahmehr*, one of the titles used for Mohammad Reza Shah (Mojtahedi, interview with HOHP). This suggestion was intended to secure greater economic resources and more administrative flexibility. As will be discussed, the way this suggestion unfolded was, however, different from what Mojtahedi intended. Given the affiliation of AMUT with the court itself, Mojtahedi had formal meetings with the Shah every fifteen days. This practice continued with all future chancellors of the university, which gave them the opportunity to discuss matters related to university with the Shah in their routine personal audiences.

Known for his pragmatism and managerial efficiency, Mojtahedi achieved the extraordinary feat of swiftly establishing a fully operational university in less than six months. In a stunning display of efficiency, the university had already opened its doors for student admissions by September 1966, leaving observers in awe of the rapid and seamless execution of Mojtahedi's visionary blueprint.

In addition to his pragmatism, Mojtahedi was also widely regarded as shrewd and skillful in his ability to secure funds not only from the Shah and the royal family but also from private donors. These private donors included former Alborz students now studying abroad and Alborz teachers who, given Mojtahedi's credibility, were willing to donate up to 10 percent of their monthly salary to help with the establishment of AMUT (Mojtahedi, interview with HOHP; also see Second Meeting of the AMUT Board of Trustees, dated 8/9/1344, NLAI-230-035171).

Beside the financial support from the court and Albroz-affiliated alumni and staff, another important measure that assisted the university financially was the agreement by which the Ministry of Oil would provide a budget to the university. This arrangement was first announced during the third meeting of the board of trustees by Dr. Manouchehr Eghbal, the chairman of the National Iranian Oil Company, who also served as the head of the board of trustees. The money initially was to support purchasing equipment for the laboratories. The support was not limited to providing money—the bureau responsible for purchases at the National Iranian Oil Company, in close relation with Abulhassan Abuzar, a member of AMUT board of trustees, also assisted in the communications with vendors and placing the orders (*Kārnāmah dah sālah fa'āliyat dānishgāh ṣan'atī* 1977, 34). Iran's Central Bank and Insurance also assisted with easing and reducing insurance costs associated with the purchase agreements.

Mojtahedi's connections with Alborz College alumni, many of whom had left the country to become respected graduates of some of the best universities in Europe and North America at the time (Zarghamee 2011), proved an invaluable asset to Mojtahedi in his new leadership role at AMUT (Partovi 2016). Firooz Partovi (b. 1936),[8] a 1963 graduate of the Physics Department at MIT and the first professor recruited at AMUT, recounts that he was approached by Mojtahedi, who informed him of his plans to establish a university modeled after MIT.

Mojtahedi embarked on two international journeys, the first in the spring of 1966 and the second in the spring of the following year. Mojtahedi had specific goals in mind. First, he aimed to reconnect with former Alborz alumni and encourage them to return to assume teaching positions at AMUT. Second, he focused on initiating negotiations with various universities to establish partnerships, or "*jumelées*," between their faculties and AMUT. Notably, AMUT was the first university in Iran where the curriculum of each faculty had to be aligned with its counterpart at a prestigious global university (*Rūzigār sharīf* 2016, 40; also see: *Nashrīyah dānishgāh ṣan'atī āryāmihr, duvvumīn sālgard ta'sīs*, 11 aban 1346, PLI-11-0620–25362).

Notably, Mojtahedi personally opposed the idea of aligning the entire university with a single partner institution. As a graduate of the Sorbonne, he was very much shaped by the European academic world. Consequently, following these trips, all the partner universities were of European origin. Among the universities paired with AMUT by Mojtahedi were Imperial College London, Technical University of Aachen, Zurich University, and the Polytechniques of Lausanne, Grenoble, and Sorbonne (dānishgāh ṣan'atī āryāmihr, duvvumīn sālgard ta'sīs, 11 aban 1346, PLI- 11–0620—25362).[9]

Mojtahedi notes that seventy-eight alumni responded positively to his invitation (Mojtahedi, interview with HOHP).[10] It is truly remarkable to contemplate the breadth of his network during that period, which enabled him to rapidly populate the new institution with highly skilled young academics. Furthermore, AMUT distinguished itself by offering competitive monthly salaries to professors holding PhD degrees (Mojtahedi, interview with HOHP). This remuneration was notably higher compared to at other universities in the capital. Mojtahedi argued that several of these professors already had positions in Europe and North America, and if they were not paid well, would likely seek employment elsewhere (Mojtahedi, interview with HOHP).

While many alumni and young scholars accepted Mojtahedi's invitation to teach at AMUT and submitted their interest, a few declined. A noteworthy example was Mostafa Chamran (1932–1981), a Muslim revolutionary who later became minister of national defense, after the revolution. In a letter to Ebrahim Yazdi (1931–2017; who later became the first minister of foreign affairs in the postrevolutionary regime), Chamran reveals some of the reasons for his ambivalence, which helps illuminate the political environment of Iran at the time, as well as the ideological debates among people of his generation:

Apart from danger and prison, our big problem is the psychological problem of working in a university whose reputation and progress would implicitly promote and would be in favor of the Shah and the relative consolidation of his position! They [i.e., SAVAK] will also monitor us and it is not known, maybe in the future they will force us to make a speech in front of the Shah and flatter him. What kind of reaction will this have on the thoughts and feelings of friends inside and outside? This is the most upsetting issue for me . . . this is where I still haven't been able to decide and convince myself, and I haven't given an answer to Dr. Mojtahedi either. (*Rūzigār sharīf* 2016, 39 [translated from Persian])

Chamran's misgivings about the new university stemmed from his perception that the Shah's authority would be absolute and carry grave consequences for dissidents of the Shah, powerfully foreshadowing the momentous episodes of repression and activism in the years that followed.

In the middle of the educational year of 1967–1968, Shah appointed Fazlullah Reza (1915–2019), a graduate of Columbia University, to replace Mojtahedi as the new chancellor, and Reza remained in this position until August 23, 1968 (*Kārnāmah dah sālah fa'āliyat dānishgāh ṣan'atī* 1977, 50). Regrettably, no records exist that provide conclusive evidence regarding the motives behind the appointments and dismissals of chancellors at AMUT. Based on multiple accounts, the Shah played a pivotal role in the removal of each AMUT chancellor, including Mojtahedi. However, according to Mojtahedi, he was never informed directly about his removal. Instead, he learned about it from a friend who informed him of the appointment of

Professor Reza as the new chancellor. Later, the prime minister also confirmed the news to him. Mojtahedi expressed his dissatisfaction about the decision and the circumstances surrounding his replacement. He mentions that his removal occurred shortly after a visit by a group associated with the US embassy in Tehran. He notes that he refused to accompany them on the tour because they had shown up without any prior arrangement. Attributing his removal from his position to this visit by US affiliates, Mojtahedi criticizes the Shah's approach. According to the perspective of Mojtahedi, although the Shah was a patriotic figure, he was susceptible to manipulation and lacked "firmness of mind (za'if al-nafs)" (Mojtahedi, interview with HOHP).

Zarghamee, who served as the fifth chancellor of AMUT, praised Mojtahedi for all his contributions but also attributed a specific shortcoming—or better to describe it as managerial attitude—to Mojtahedi. This perceived fault pertains to Mojtahedi's approach to discipline, characterized by his personal monitoring of student and faculty attendance through note taking. According to Zarghamee, such an approach made Mojtahedi's mode of management not suitable for a university. This observation about Mojtahedi's management style has been corroborated by others, including Hossein Amanat, a former alumnus of Alborz and one of the early architects of AMUT whom we also interviewed directly for this project (interview with Amanat, 2021).

In sum, Mojtahedi's management methods, though more applicable for colleges and high schools, were foundational in shaping AMUT's early framework. Despite his controversial departure, his legacy as an efficient and visionary founder continues to influence the university's culture and policies. His commitment to a highly competitive student body and a patriotic and Iran-oriented educational philosophy, although sometimes at odds with external and political expectations, established a distinct identity for AMUT that endures to this day.

EDUCATIONAL ORGANIZATION OF AMUT (1966–1979)

AMUT opened in September 1966 with fifty-four faculty members. During the recruitment process, 125 Iranian graduates of foreign universities submitted their interest, and thirty-six were selected (including eight PhD graduates, twenty master graduates, three bachelor graduates). In addition, eighteen Iranian master graduates of STEM education were also hired to work in laboratories (*Kārnāmah dah sālah fa'āliyat dānishgāh ṣan'atī* 1977, 7). Since its first year of operation, the students of this new campus have been highly capable. Partovi notes that the students who participated in the entrance exam for AMUT during its inaugural year were of very high caliber. He attributes such interest in the university to the students' familiarity with Mojtahedi from his role as the manager of the highly esteemed Alborz College (Partovi 2016). As for enrollment, in the first year, there were a total of 407 students, thirteen of which were female (*Kārnāmah dah sālah fa'āliyat dānishgāh ṣan'atī* 1977, 7). During its initial year, the university had six faculties: Electrical Engineering (specializing in electronics and electrotechnics), Mechanical Engineering (with two specializations: machinery and installations), Chemical Engineering, Metallurgy Engineering, Civil Engineering (covering mathematics, physics, and chemistry), and Industrial Management and Engineering Economics (*Rūzigār sharīf* 2016, 46). As for library development, the Tehran branch of the Franklin Book Programs, a US not-for-profit organization of the Cold War era aiming to make American books available in "third world" countries, contributed $100,000 (Ganjavi 2023), and the British Council provided 489 books (*Rūzigār sharīf* 2016, 61).

During its second year of operation, the number of departments at AMUT expanded to include the Department of Physics and Mathematics and the Department of Industrial Management and Engineering Economics. In addition, the academic staff grew to seventy individuals.[11]

Also, the educational system of the university changed to a course-based one. Previously, all students had to take some general courses before enrolling in any course in their respective department (*Kārnāmah dah sālah fa'āliyat dānishgāh ṣan'atī* 1977, 50). As part of this shift in the university's educational system, each course became independent of other courses, meaning that once a student passed a course, they would not have to take it again. Also, each student would have started their educational journey in their respective department.[12]

In August 1968, Mohammad Reza Amin (1927–2003), who earned a PhD in physics from the University of California, Berkeley, and at the time was the deputy of the Bank of Industrial and Mining Development of Iran, replaced Dr. Reza as chancellor (rūznmah rasmī kishvar shāhanshāhī, 21 Aban 1347, 8, NLAI-230-035171). He was chancellor for four years. Under the leadership of Amin, the university experienced a sense of order and discipline in both its educational and administrative aspects (*Rūzigār sharīf* 2016; also interview with Partovi, 2024). The university council compiled a set of educational regulations on May 20, 1969, which consisted of fifty-two articles and thirty notes. These regulations were approved by the board of trustees of the university in the same year and subsequently implemented.

During this period, Chancellor Amin implemented additional measures, including transforming professors into full-time positions. The responsibility for educational matters was handed over to the university council, consisting of the university chancellor, department heads, and elected representatives from each of the department councils. An appointed committee of professors, known as Welfare Committee (*Kumiteh rifāh*), was tasked with evaluating and enhancing the teaching staff, considering factors such as educational quality and quantity as well as their societal contributions.[13] Various educational rules, which required the expulsion of students failing to meet academic standards, were put into effect (NLAI-230-024515).

In addition, during the academic year 1968–1969, the Faculty of Science Engineering was divided into three separate faculties: mathematics, physics, and chemistry. The students were allocated to these newly established faculties accordingly. Furthermore, the Faculty of Industrial Management and Engineering Economics now admitted undergraduate students for the first time. Of the 263 students who were among the first graduates of the AMUT (graduated in 1969–1970), 101 started working in Isfahan Steel Company, 25 started working at AMUT, and several others in Iran's television stations, other universities, and regional electricity organizations (*Kārnāmah dah sālah fa'āliyat dānishgāh ṣan'atī* 1977, 169).

In 1972, Dr. Seyyed Hossein Nasr replaced Dr. Amin. During his tenure, a few important institutional developments occurred: First, the affiliation between MIT and AMUT was fortified; second, Center for Humanities,[14] a center offering elective courses in the field of humanities, was established at AMUT.[15] Beginning in the academic year 1973–1974, the Department of Computing and Computing Applications was merged with the School of Mathematics, resulting in the creation of the School of Mathematics and Computer Science. Postgraduate courses in the fields of mathematics and computer science were introduced, and students were accepted for the first time. Furthermore, the university council approved the establishment of two units for sports and physical education, which became compulsory courses for all undergraduate students starting in September 1973. It was also in this year that the first postgraduate student was admitted to the field of chemical engineering. Moreover, the field of metallurgy was introduced as a major for the first time in this educational year (*Sharīf az āqāz tā kunūn bi rivāyat 'asātīd* 2009, 116).

Ten years after its establishment, AMUT's student enrollment of 407 in the first year had increased to 2,942, and the academic staff had increased from 54 to 368 professors, assistant professors, instructors, and teaching assistants (*Kārnāmah dah sālah fa'āliyat dānishgāh ṣan'atī*, 1977, 7). According to a report by Iran's Ministry of Science and Higher Education, in its tenth year of operation, researchers at AMUT were in charge of more than 62.5 percent of the research projects in the field of engineering in Iran (*Kārnāmah dah sālah fa'āliyat dānishgāh ṣan'atī*, 1977, 7). The university had developed to one with nine departments (Mechanical Engineering, Math, Computer Science, Structural Engineering, Metallurgy, Physics, Chemical Engineering, Industrial Engineering, and Electrical Engineering) and ten affiliated educational and research institutions. By 1976, one-fourth of all engineers in Iran were graduates of AMUT (Mohammad Ali Ranjbar interviewed in *Sharīf az āqāz tā kunūn bi rivāyat ru'asāy-i ān* 2006, 40).

Before ending this section, some statistical information on student enrollment and gender and social diversity is important, not only to provide a sense of the student body but also for our examination of political activity and gender relations in the following chapters. As for enrollment, although the first year had a total of 407 students, 13 of which were female, the second year did not see a rise in female students: From a total of 368 students, only 4 were women (*Kārnāmah dah sālah fa'āliyat dānishgāh ṣan'atī* 1977, 53). In the third year, the total number of students increased to 413, 28 of which were female. (*Kārnāmah dah sālah fa'āliyat dānishgāh ṣan'atī* 1977, 72). In 1969–1970, the total number of students was 370, with 15 female students. (*Kārnāmah dah sālah fa'āliyat dānishgāh ṣan'atī* 1977, 93, 1977). In 1970–1971, the total number increased to 554, with 44 female students (*Kārnāmah dah sālah fa'āliyat dānishgāh ṣan'atī* 1977, 140). In the sixth year of its operation (1971–1972), 566 students were accepted, 44 of which were female (*Kārnāmah dah sālah fa'āliyat dānishgāh ṣan'atī* 1977, 197). In the seventh year, of the total 637 newly admitted students, 63 were female (*Kārnāmah dah sālah fa'āliyat dānishgāh ṣan'atī* 1977, 255). Next year, the total was 632, with 77 female students among them (*Kārnāmah dah sālah fa'āliyat dānishgāh ṣan'atī* 1977, 341) . In the ninth year, the total students were 510, out of which 41 were female (*Kārnāmah dah sālah fa'āliyat dānishgāh ṣan'atī* 1977, 417). Also, AMUT started offering graduate courses (a master's program) in 1971, starting with twenty-three, of which two were female. By 1975, this had increased to seventy-nine students, with five females among them (*Kārnāmah dah sālah fa'āliyat dānishgāh ṣan'atī* 1977, 440).

It is difficult to obtain an exact understanding of the class and social conditions of the student attendees at AMUT during those years. We know that in 1971–1972, 49.7 percent of the student acceptances were from Tehran, 9.9 percent from Isfahan, 5.1 percent from Khorasan, 4.2 percent from Eastern Azerbaijan, 4 percent from Gilan, 3.8 percent from Mazandaran, 2.8 percent from Fars, and the remainder from other places in the country, with only 0.2 percent from the Baluchi areas of Iran, and a total of less than 2 percent from Kurdish areas of Iran (*Pīshnahād darabārah falsafah 'āmuzish va siyāsat-hā-yi dānishgāh ṣan'atī āryāmihr dar iṣfahan*, 1351, NLAI-230-027900). This shows, first, that at least up to this time and probably in the following years that fall within our study period, there was a sharp imbalance between attendees from the capital and other areas, and also that the majority of attendees were from Persian and central areas of Iran (around 70 percent). Still, given the fact that people living in Tehran might be from various ethnic populations, especially of Turkish ethnicity, the previous percentage of 70 percent in terms of ethnicity should be considered just as an informed guess. The same report also notes that of the total, 24.3 percent of students came from families with a monthly income of 5,000 rials, 22.4 percent with 5,000 to 7,000 rials, 21.2 percent with 7,500 to 11,000 rials, 17.7 percent with 11,000 to 26,000 rials, and 14.5 percent with more than 26,000 rials (*Pīshnahād darabārah falsafah 'āmuzish va siyāsat-hā-yi dānishgāh ṣan'atī āryāmihr dar iṣfahan*, 1351, NLAI-230-027900).

While we were unable to locate definitive archival records, it is important to note that AMUT students largely came from lower-middle-class and moderately religious families. This was true of the individuals we interviewed and also reflected in formal statements by university leaders. For example, in an interview conducted with Chancellor Nasr by Hossein Ziya'i in October 1982–January 1983 (by Foundation for Iranian Studies [FIS]), Nasr describes the student demographics in this way:

The population of our universities were growing, and the wealthier the society was getting the upper class strata of the society that did not have strong religious beliefs anymore was sending their kids abroad for education. The kids of lower classes with stronger religious beliefs studied hard and got into universities like AMUT, they did not have the means to send their kids to Europe and the US . . . in 1971 onwards the number of students from poor and religious backgrounds who came to universities kept growing. (Nasr, interview with FIS, translated from Persian)

In addition, Chancellor Mehdi Zarghamee has described the social class of students at AMUT as follows:

The social class of the students is like this: approximately 50% of the students were Tehrani, and 50% came from other towns. Some of the provinces never contributed much, except one or two, to Aryamehr University. Aryamehr University was taking the best of technically prone students. I wasn't amazed that Sistan and Baluchistan bander <ports> were provinces that never contributed. But then, later on, I found out that Kermanshah wasn't doing so well, either. I mean, the point was this: some of the wealthy provinces even were amongst those that never contributed. (Zarghamee, interview with HOHP)

Zarghamee further notes that a large group of the students who were not from Tehran came from Mashhad and Isfahan, two of the main Persian-speaking provinces of the country. In Iran, these cities are known to skew toward more traditional and religious families. Students from these two provinces, along with a third Persian-speaking province, Kerman, were also more politically active.[16] In terms of economic class and background, approximately 5 percent of students came from a "professional" background, with 30 percent from families of "white-collar employees," and the rest (65 percent) from "blue-collar" families (Zarghamee, interview with HOHP).

RESEARCH PRIORITIES AND INDUSTRY CONNECTIONS

Research areas that were prominent from early days include water shortage and water desalination (as a result of Iran's water shortage problems), plasma, and the realm of communication and biophysical communications. In a few years, the university further developed its relations with the industry and made agreements with Iran's navy, Planning Organization, the Organization for Regional Water, Organization for Standards, Iran's National Steel Company, and some industrial and commercial sectors (*Kārnāmah dah sālah fa'āliyat dānishgāh ṣan'atī* 1977, 116).

AMUT also played an important role in the Isfahan Symposium on Laser, which occurred during the chancellorship of Dr. M. Reza Amin in 1971. Seventy-five non-Iranian researchers, four of whom were Nobel prize winners, participated in this conference on laser physics, along with twenty Iranian researchers (*Kārnāmah dah sālah fa'āliyat dānishgāh ṣan'atī* 1977, 105). Ali Javan (1926–2016), from MIT, played an important role in leading this conference. Aleksandr Mikhailovich Prokhorov (Nobel Prize winner in physics, 1964), Charles Hard Townes (Nobel Prize winner in physics, 1964, divided with Prokhorov), Gerhard Herzberg (Nobel Prize winner in chemistry, 1971) were among the attendees.[17]

A closer archival investigation into some of the early AMUT research projects would help shed some light the complexities and reasonings behind such initiatives. An early example is the trip to Germany and France in 1967 by Nusrat Allah Vahidi and Ali Payman to discuss possible collaborations in the field of space research. This trip was financially supported by the office of the prime minister and made following a confidential letter by the prime minister (Amir Abbas Hoveyda) requesting that the professors explore prospects for a project to build long-range missiles for space research and an Iranian-German satellite for telephone and television communication. The project also aimed to train technical personnel for these objectives. The AMUT professors returned from the trip and provided a chronology of how such a proposed program could achieve its goals (NLAI-230-035171). This trip was soon criticized by the chancellor of the AMUT, Mohammad Ali Mojtahedi, who, in a confidential letter to the office of prime minister (dated March 16, 1967), objected and insisted that the scheduling of the trip needed to be done in a way not to interfere with academic and teaching obligations of the professors involved (NLAI-230-035171).

Another example of research initiatives is a report from an AMUT committee (dated July 5, 1967) to the prime minister of Iran. This report describes the various initiatives and endeavors undertaken by AMUT

1.3 Iran's Shah greeting professors attending the Esfehan Laser Symposium, 1971. Right to left: Firooz Partovi, Aleksandr Mikhailovich Prokhorov, unknown, Gerhard Herzberg, Nicolaas Bloembergen, unknown, Charles Townes, unknown. Photo credit: Personal collection of Firooz Partovi.

faculty for implementing a water desalination project, which was given support and encouragement by the Iranian prime minister two months earlier. The group reports how the team tested the water in the southern coastal regions Iran, Qishm and Bandarabbad, and then started a collaboration with several West German companies and research institutions in the field, including the Geesthacht (West German) Nuclear Research Center (NLAI-230-035171).

In 1971, a radioactivity laboratory was equipped through the financial support of the Iranian government. The objective of this laboratory was to work in the field of industrial usage of radioactive substances. The laboratory conducted numerous different projects, including the designing of a nuclear seismometer (1972), as well as research into the building of biological filters and work to facilitate the industrial usage of lasers in Iran. Other projects included making ionisers and multi-end detectors and research on using radioactive substances in the oil industry (*būdjah sāl 1353 markaz kārburd mavād rādiyuactiv dar ṣan'at*, NLAI-230-024244). To provide further insights into the history of research at AMUT, some reflections from Mehdi Zarghamee, a chancellor of the university, prove insightful. He notes:

I remember at Aryamehr University we started a research was on [sic] coding/ decoding devices. We formed a laboratory which was called Electronics Research Lab. And we developed an instrument that could decode all the vocal coded things. . . . The coding and decoding devices that Iran [sic] had were very mundane, elementary kinds that Iran [sic] bought from Germany . . . anyway we called Edareh-Dovvom [Second Department] in the army. (Zarghamee, interview with HOHP)

As the story unfolds, Zarghamee relays that the army tested the decoding system, saw its capabilities, and decided it was something of value to them. The Shah asked the chiefs of staff to sponsor research "and they actually came up with a proposal to make a whole research center in Larizan [sic] for coding/decoding electronic devices that we would develop" (Zarghamee, interview with HOHP).

In another case of presenting research to the Shah, Zarghamee discusses the research on increasing the efficiency of the Paykan engine. Paykan[18] was Iran's first domestically produced automobile. Over the course of several decades, Paykan maintained its status as an economical option accessible to the broader populace. Zarghamee notes that in his audiences with the Shah, he used to give him "a flavor of what research really is going to do, and how it should be directed." Zarghamee points out that the reason for this had been to make the importance of research at universities clear to him.

Because the Shah was—in 1968 there was a Ramsar conference, and the Shah was . . . against research. In 1969, he had accepted that there would be research. He thought that research would be a waste of time in 1968. At that time he wasn't really convinced that research would be that useful . . . His commitment to research was—I mean, it was very hard to convince him that, really, technical research was meaningful or was necessary. He wouldn't give priority to that. (Zarghamee, interview with HOHP)

Archival materials provide insight into the pivotal role of AMUT in various research projects in Iran. From space research collaborations to water desalination initiatives and nuclear applications, AMUT's contributions spanned a diverse range of fields. The reflections by Mehdi Zarghamee further underscores the evolution of the Shah's perspective on research, indicating a transition from initial doubt to eventual recognition of its value. As AMUT evolved from a teaching to a research institution, the university also took on a new global significance. There were growing concerns internationally regarding the Shah's ambitions to transform Iran into a science and technology powerhouse in the Middle East. This is best exemplified by reactions to the new university in the United States, and, in particular, within MIT.

2

THE MIT–AMUT PARTNERSHIP IN
SHAPING A REGIONAL STEM LEADER

In the current geopolitical context, few can remember a time when the US and Iran had amicable relations. In the American imagination, Iran commonly conjures images of the hostage crisis (1979–1981), the eight-year Iran-Iraq war (1980–1988), enmity toward Israel, state sponsorship of terrorism, and a brutal Islamic dictatorship. Yet the historical record contradicts, or at least complicates, these images. The Shah of Iran was a close American ally, and many have argued that before the 1979 revolution, Iran was the most important strategic partner of the US in the Middle East. In this chapter, we zoom in to these relations through the prism of an elite engineering institution, Aryamehr University of Technology (AMUT). In many ways, the story of AMUT is an untold story of US-Iran cooperation, a moment in global affairs where Iran and the United States entered a contractual partnership to develop the most prominent STEM university in the region. Ironically, that same institution played a nontrivial role in the 1979 revolution, which has defined US-Iran relations since. The circumstances, dynamics, and lessons from the Iran-US cooperation to design AMUT are the focus of this chapter.

HISTORICAL CONTEXT OF US-IRAN ACADEMIC COOPERATIONS

Following the CIA-backed overthrow of Prime Minister Mohammad Mosaddegh in 1953 and the reinstatement of Mohammad Reza Shah Pahlavi, diplomatic relations between the US and Iran resumed while the Cold War intensified. Educational reform emerged as a key policy priority for the US government during this era. Domestically, the 1958 National Defense Education Act (NDEA), Title IV of the Higher Education Act in 1965, and the Elementary and Secondary Education Act (1965) bolstered American K–12 and higher education. Internationally, US foreign assistance programs provided technical training, curriculum development, and infrastructure and improvement projects abroad. Education, seen as a powerful tool of soft power capable of influencing nations and winning hearts and minds, emerged as a means for the US to foster amicable relations with countries vulnerable to Soviet influence (Shannon 2017).[1]

During the 1950s and 1960s, numerous American universities eagerly supported the US government's educational aid programs abroad. A notable example is Brigham Young University (BYU), which actively participated in the modernization of Iranian education. From 1951 to 1955, BYU collaborated with the Iranian Ministry of Education to restructure the curriculum and dispatched specialists to assist in training elementary and secondary school teachers (Garlitz 2012). Utah State Agricultural College (USAC) worked closely with the agricultural faculty of the University of Tehran (Karaj College) to modernize its operations, based on the model of US agricultural colleges. This affiliation with Iranian agricultural educational programs continued until 1964, with a focus on spreading modern agricultural techniques to rural areas. In addition, the University of Pennsylvania sent advisors to Shiraz with the aim of transforming Pahlavi University to align with the US higher education system from 1962 to 1967 (Garlitz 2012; Ganjavi 2023). The Massachusetts Institude of Technology's (MIT's) involvement with Iran also can be traced to this era, reflecting a broader trend of international collaboration between institutions and governments.

The collaborations between AMUT and MIT started in the early 1970s, during the chancellorship of Dr. Reza Amin at AMUT, with the Isfahan Symposium on Laser, discussed in the previous chapter, in which Ali Javan, Iranian-American professor of physics at MIT, played an important role. Also, Gordon Brown, dean of MIT School of Engineering (1959–1968), was foundational in further fostering these connections. In June 1971, Brown, accompanied by advisors from the international consulting firm Arthur D.

Little, arrived in Iran to develop a comprehensive plan for the AMUT campus in Isfahan (*Rūzigār sharīf* 2016, 121).

While the early collaborations started during the tenure of Reza Amin, it was during the chancellorship of his successor, Dr. Seyyed Hossein Nasr, that the relationship between the two institutions was contractually formulated and extended to several new areas. We will discuss the tenure of Dr. Nasr, his views, and his legacy in detail in the next chapter. Here, we will mainly focus on understanding the arrangements between the two campuses and how MIT students and faculty reacted to some of those partnerships.

MIT AND AMUT ENTER INTO A CONTRACTUAL PARTNERSHIP

On June 20, 1974, MIT President Jerome Wiesner (1915–1994) reported to the Academic Council, J. R. Killian (former president of MIT), H. W. Johnson (also former president of MIT), and Gordon Brown regarding his trip to Iran in June 1974. He discussed the high-profile people he met, including the Shah. He also talked about the various interests that people had expressed, including chemical engineering, as well as the social sciences, including both economics and issues of science and technology in a contemporary society. Wiesner presented the Shah with a fragment of a crystal from one of the crystals grown aboard Skylab 3 and 4 by the Skylab astronauts (MIT Archives: Box 90 [Barcode: 39080020389232], HD [HD], MC—0420, Jerome B. Wiesner papers). This special and uncommon gift was intended to build trust and symbolize MIT's significant scientific contributions, hoping to cement the Shah's commitment to the partnership between the two institutions. As we will see, Wiesner soon officially invited the Shah to visit MIT.

In June 1974, the revised agreement for a program of collaboration between MIT and AMUT was prepared. This document discussed the objectives of this program, its content, procedures, budget, method of payment, and duration. The introduction to this agreement reads:

The mandate set forth by His Imperial Majesty the Shah for the Arya-Mehr University of Technology (AMUT) is, on the one hand, to educate a group of elite engineers and scientists who will become the key instruments of the future economic and social development of Iran and, on the other hand, through research

2.1 From right to left: Firooz Partovi, Mehdi Zarghamee, Jerome Wiesner, Hossein Nasr, Mokhber, Mostafa Toosi, May 1974. Photo credit: Personal collection of Firooz Partovi

and development programs conducted on the campus by students and faculty, to accelerate the transfer of science and technology into the societal fabric of Iran to ameliorate the pressing industrial, economic, social, and human problems of a fast-paced industrializing society.

The agreement also encompassed some details regarding the historical development and future prospects of AMUT:

AMUT is presently operating an eight-year-old campus of about 3,000 students and 300 faculty members in Tehran—the average age of the faculty is about thirty. The Arya-Mehr University of Technology at Isfahan is expected to start classes in September 1976. The Isfahan campus will have an innovative program which has been developed with the assistance of Professor Gordon S. Brown of MIT and is expected to have 8–10 thousand students and about eleven hundred faculty in the final stages of development.[2]

An important aspect of the partnership was the inclusion of an agreement for Iranian professors to undertake residencies at MIT, fostering an exchange of knowledge and expertise that would enhance the academic capabilities at AMUT. The details of this arrangement reveal the significance of the role of MIT as a model to be emulated by AMUT faculty:

To enable the present and future members of the AMUT faculty, through a period of residence at MIT, to be introduced to the frontiers of science and technology in their fields of interest; to obtain an awareness of the spirit of MIT; to understand how MIT organizes and operates to innovate change; to acquire a first-hand experience with the mechanisms that have contributed to the success of the faculty, students, and administration of MIT, in adapting to challenges set forth by a changing global society.[3]

According to this agreement, the intention of the program was not that the development of AMUT would be determined by MIT, but, rather, through this exchange, AMUT would be able to establish its own educational and research programs on a firm course compatible with the culture and problems of Iran, and through the strength derived from the experiences of AMUT faculty at MIT, to take long-range steps with confidence.

The program provided the following:

1) For the training, leading to advanced degrees, at MIT of a selected number of outstanding students who are expected to be employed as members of the faculty of AMUT at Isfahan.
2) For selected members of the faculty of AMUT to spend not less than one semester as visiting faculty at MIT.
3) For a limited number of administrative personnel—librarians, admissions office staff, research administrative staff, and so on, to spend a period of residence at MIT.
4) For the faculty and staff of MIT to develop working professional ties with the faculty of AMUT through visits by MIT faculty and administrators to Iran, and vice versa, and by collaborative research projects of mutual interest that are particularly relevant to the problems of Iran.[4]

It was expected that the activities embraced by this program would build gradually to the level where six to eight graduate students and three to six visiting faculty members or administrative staff would complete residencies at MIT. In this agreement, it was estimated that between four and six graduate students would be in residence during the academic year 1974–1975, and possibly two or three faculty visitors.

This agreement was signed for the period July 1, 1974, to June 30, 1979.[5] The total estimated cost of this program for AMUT was $27,000. Title to any invention or discovery made or conceived in the performance of research at MIT would have remained with MIT, which would have the sole right to determine the disposition of any patents or other rights resulting from there, provided that, upon issue of any patent on any such invention or discovery, MIT would have granted to AMUT an irrevocable, royalty-free, nonexclusive license for use of such invention or discovery for its own purposes.[6] While, as noted earlier, the agreement made specific provisions concerning the training of future professors at AMUT in Isfahan, visiting faculty at MIT, and the development of working professional ties between the faculty and staff of MIT and AMUT, as well as issues regarding student intellectual property and financial commitments between the universities, the scope of collaboration between the two institutions extended significantly beyond these specific arrangements. Indeed, the collaboration ventured into expansive fields such as nuclear engineering and oceanography, among others, reflecting a multifaceted partnership aimed at fostering extensive academic and scientific exchanges. This broader engagement was further highlighted when on July 19, 1974, a letter was sent from Kazimiyan, the minister counsellor for cultural affairs, to Professor Edward A. Mason, chairman of the Nuclear Engineering Department at MIT. The letter informed Professor Mason about the Shah's initiation of a program focused on nuclear science and technology.[7]

On July 23, 1974, Professor Jerome B. Wiesner, president of MIT, wrote a letter to invite the Shah for a visit to the MIT campus. In this letter, Wiesner referred to a conversation he had with the Shah during his visit in Tehran in which he told the Shah, "MIT has developed its strength and usefulness to our society in large measure because of the close relationship we have always had with the practical men who were building this country." He encouraged the Shah to be accompanied in his visit to MIT by "practical people who are building the country."[8]

Three weeks later, on August 12, 1974, the Shah replied to Wiesner, saying that he was interested in visiting the campus and planned to do so whenever he had the opportunity of coming to the US. He further noted:

In view of the distinguished record of MIT in providing excellent academic training, I am convinced that the exchange of students and faculty between the Aryamehr University of Technology and your Institute and practical men from both countries would be of enhanced mutual value and would also be conducive to furthering the educational goals we have set ourselves in Iran.[9]

The MIT correspondences extended to Iran's ministers, such as Iran's minister of science and higher education, Abdol Hossein Sami'i. On August 22, 1974, Wiesner sent a letter expanding on his previous suggestion in July 1974 regarding possible cooperation in the field of oceanography. It seems he had already written about such a possible cooperation, received a letter of interest from Iran, and thus was following up to provide additional details.[10] Wiesner informed Minister Hossein Sami'i of the need for a properly equipped ship for the program, which could cost approximately six million dollars. However, he informed Sami'i of a boat named *The Chain*, an old but sound vessel that could be purchased for $2 million and modernized for this purpose.[11]

On September 19, 1974, Chancellor Nasr wrote a letter to President Wiesner, noting the interest in Iran in research on lasers and articulating that soon either AMUT or a joint national committee would be in touch with Ali Javan on this matter. In this correspondence, Nasr also conveyed that a new research center had been established at AMUT, modeled after the Lincoln Laboratories at MIT, and its head, Dr. Vojdani, would be planning a visit to MIT.[12]

While the partnerships between AMUT and MIT were developing, both campuses experienced significant student activism, albeit in distinct contexts influenced by their unique historical and political orientations. The environment of MIT was particularly vibrant during the 1960s and 1970s, a period marked by significant engagement with both national and international issues. The active political context at MIT sometimes resulted in backlash against international collaborations, particularly those involving nations like Iran, which were seen through the prism of US foreign policy.

ACTIVISM AND TENSIONS AT MIT AROUND US INVOLVEMENT IN AMUT

During the 1960s and 1970s, MIT, like several other universities across the United States, was a vibrant arena of social and political activity, deeply influenced by several major historical events and movements. The Vietnam

War was a dominant issue of the time, inciting widespread protests across college campuses. Students and faculty at MIT were actively involved in demonstrations against the war. These protests were part of a larger national student movement that questioned the US role in Vietnam and criticized the military-industrial complex, of which MIT was a significant part, due to its research contracts and partnerships with the Department of Defense. President Johnson's escalation of the Vietnam War and President Nixon's subsequent administration, marred by the Watergate scandal, fueled distrust in governmental institutions and spurred further activism among the student body. MIT was also influenced by the civil rights movement, which sought to end racial segregation and discrimination against African Americans and prompted universities to reflect on their own policies and cultures. The campus itself was a microcosm of broader societal debates. Debates over free speech, academic freedom, and the responsibility of intellectuals in society were commonplace. MIT's role as a leading research institution also placed it at the heart of discussions about the ethical dimensions of scientific research, particularly in relation to war and militarism.[13]

The Science Action Coordinating Committee (SACC) was a student organization formed at MIT in 1969 with the aim of advocating for radical reform in how science and technology were utilized and applied in society (Renehan 2007, 51). Noam Chomsky played an advisory role in the creation of SACC, which quickly became a vocal critic of the scientific establishment's role in global politics. This critique particularly focused on how science and technology were implicated in broader global dynamics, such as the Vietnam War. SACC members actively challenged the ethics of using scientific advancements in warfare and other areas of conflict. Notably, their criticism of science and technology extended to specific international involvements of the US. By 1975, SACC members had voiced their concerns regarding MIT's contracts with the Iran and Taiwan programs, expressing criticism toward both (Renehan 2007, 53). Their objections highlighted the ethical dilemmas and potential political repercussions associated with these international collaborations, reflecting a broader concern about the impact of scientific work beyond academic and national boundaries.

On March 7, 1975, a student journalist from *The Tech* newspaper published an editorial titled "Selling MIT, Bombs for the Shah," revealing the details of the association between MIT and AMUT in nuclear engineering programs (*The Tech* March 7, 1975, 4). The article urged the campus community to discontinue its partnership with what was perceived as a corrupt and authoritarian Iranian government:

Iran sits in a position critical to the peace of the world today. Overlooking the powder-keg in the Mideast from near the borders of the Soviet Union, Iran, with its recent military build-up, cannot be ignored by anyone concerned with the fate of that troubled region. . . . For MIT to help introduce nuclear technology into this situation is criminal. Arms control experts have highlighted Iran as a country to be feared, reflecting the feeling that spread of nuclear arms into that region could have grave consequences for the peace of the world. MIT, of course, is not training engineers to build bombs, but the leap from reactors to weaponry is not great. (*The Tech* March 7, 1975, 4)

In this context of critical evaluation of the military spending of Iran's Shah, the article continues its argument and criticizes the behind-the-scenes negotiations between AMUT and MIT:

Given the reputation of the MIT faculty and administration, a through consideration of the moral questions and long-term implications of such a move might have been expected to win out over the increased income which will be generated for the institute by the decision. Yet discussion of the issues—nuclear proliferation, MIT's obligation to society, MIT's obligation to its own educational standards in developing special programs—has not occured within the community, as the administration moved to make the deal with Iran a *fait accompli* before the rest of the community even knew about it. (*The Tech* March 7, 1975, 4)

Another MIT student journal, *Thursday*, followed the story and published a series of articles on the topic. Later in March, during a meeting at MIT attended by approximately 500 students and faculty, President Wiesner supported this program and noted, "We've long been educating students from underdeveloped countries. Iran can expand on a more rapid basis now, and so we are expanding their programs accordingly" (quoted in *The Tech* March 18, 1975, 3). The audience, however, was highly critical and "condemned the government of Iran as dictatorial and repressive, and charged MIT with complicity in helping Iran obtain nuclear technology."

Some Iranian students at MIT also played a role in mobilizing against this partnership, and one of the students from the MIT Association of Iranian Students organization said, "The Shah of Iran Wants to use nuclear power so he gets to stop liberation movements anywhere in the Gulf area."

However, a visiting associate professor of mechanical engineering, Parviz Payvar, the dean of the Energy Division at AMUT in Iran, publicly supported the MIT program aimed at training Iranian nuclear engineers, arguing, "Iran had never been an aggressor nation in world affairs, and that, as a signator to the Nuclear Non-Proliferation Treaty, Iran would not develop nuclear weapons with the technology they were getting" (*The Tech* March 18, 1975, 3).

On March 28, 1975, Gordon Brown, who played a key role in negotiating the program, wrote a memo to Joseph Weizenbaum, a vocal critic of the program.[14] In the memo, Brown expressed his belief that refusing to train the Iranian students at MIT would be counterproductive and put the institution at a significant disadvantage. He argued that even if MIT declined to participate, Iran would still obtain nuclear power reactors from other countries and train their engineers elsewhere. Brown believed that by participating, MIT could have a role in establishing nuclear fission power in Iran and contributing to its safe and effective operation, as well as potentially developing alternative solutions in the future.[15] Brown writes:

Because I respect the integrity and value system of our faculty, I am relieved to learn that we will have a chance to instill our value system into the minds of the Iranian students, to impress upon them our concern for the establishment of strict control procedures, to give them the resolve to see it that nuclear technology is only used for peaceful purposes. . . . But I believe that if we take the position that MIT should not train these Iranians we will be taking a counterproductive action that will be greatly to our disadvantage. . . . Iran can and will obtain nuclear power reactors from other countries.[16]

In his response dated April 1, 1975, Weizenbaum countered the argument by stating that the fact that Iranians may acquire nuclear reactor technology from other sources does not justify disregarding MIT's duty to act with prudence and wisdom after careful deliberation. He suggested that if MIT were to demonstrate caution, even if it went against its immediate financial interests, it could potentially inspire others to reconsider their own actions. Weizenbaum emphasized that if everyone adopted the mindset of "if I don't do it, someone else will," this would perpetuate dangerous and unethical endeavors. Weizenbaum elaborated:

Our citizenship in the Institute community places responsibility on our shoulders for whatever important political or moral commitments the institute makes. We are accountable for their consequences. Responsibility, accountability, and authority are inseparable. As we cannot abdicate to our administrators our own responsibility and accountability in matters of conscience, so can we not abdicate to them ou authority to participate in deciding matters of such great consequences.[17]

During April 1975, significant developments arose regarding MIT's proposal to continue supporting the program. The faculty held a meeting to discuss the matter and deliberated on a motion put forward by faculty peace activist Philip Morrison to terminate MIT's backing for the program. In the same month, the students expressed their opposition to the plan through a vote on the question during the Undergraduate Association elections, with the result being 1,001 against and 214 in favor.

On April 15, 1975, an article written by Kent F. Hansen titled "MIT Should Help Fulfill Industrialization Goals" was published in *The Tech*, specifically discussing the institute's role in supporting Iranian industrialization efforts. In the same issue, Weizenbaum published an essay titled "Identification with Iran Identifies Us with Torture." The debate on the program persisted, and it was further fueled by an article titled "A Bomb for Iran 'Unfeasible'?" written by Thomas J. Spisak and published in *The Tech* on April 25, 1975.

On April 25, 1975, a sit-in protest against the Iran deal was organized. Approximately 200 individuals participated in the rally, during which a symbolic atomic missile was placed in front of the student center as a visual political statement. The protest movement was led by a newly formed group called the Coalition against Training Nuclear Engineers for the Shah (CATNES), which co-organized it along with SACC (*The Tech*, April 29, 1975). Moreover, a series of teach-ins focused on the Iran program drew crowds to one of the largest meeting rooms on campus. These events sparked significant engagement, with 1,200 individuals signing a petition expressing their opposition to the program (Renehan 2007, 64). In addition, an undergraduate student resolution was passed, condemning the decision. In further protest, students organized a sit-in within the Nuclear Engineering Department during the same month.

On May 7, 1975, Professor Charles P. Kindleberger, an economics professor, was appointed as the leader of an ad hoc committee consisting of both students and faculty members. The purpose of this committee was to examine MIT's international institutional commitments, specifically in light of the month-long debate surrounding the Iranian program. The decision to establish this committee was made by the faculty (*The Tech*, May 9, 1975).

On November 21, 1975, *The Tech* published an article providing a detailed discussion of the findings and work conducted by the ad hoc student-faculty committee. After seven months of work, this group proposed a "standing committee with broad powers to investigate and review international research, education, and service agreements"; however, the faculty overrode the proposal and substituted the standing committee with a "temporary group with curtailed authority to review projects for a year before reporting again to the faculty" (*The Tech*, November 21, 1975).

By 1976, discussions occurred about scaling back MIT projects in Tehran due to political and civil unrest in Iran.[18] President Wiesner discussed MIT's international projects with Gordon Brown, particularly focusing on the increasing turmoil in Iran. On April 14, 1978, an Iranian student from MIT caused a disruption during a speech given by former CIA director William Colby on campus (*The Tech*, April 14, 1978).[19]

On October 15, 1978, a teach-in addressing the Iranian dissension took place on the MIT campus. Notably, Nobel Laureates from MIT and Harvard delivered speeches during the event. A report written by Elaine Douglas titled "Iranians Charge Media Cover-Up" appeared in *The Tech* newspaper on October 17, 1978.

SHIFTING ALLIANCES AND EDUCATIONAL DIPLOMACY: THE COLD WAR, MIT, AND IRANIAN RELATIONS

The educational relations between the US and Iran underwent significant shifts during the Cold War era, primarily driven by the geopolitical strategies of the two nations. After the overthrow of Prime Minister Mohammad Mosaddegh in 1953, and the subsequent changes in the Iranian political landscape, educational assistance programs from the US gained momentum. Education was recognized as a potent tool for soft power, fostering friendly ties with nations thought to be vulnerable to Soviet influence. The collaboration focused on military training, technical education, and cultural exchange, with Iranian students traveling to the US for studies. Amid these developments, MIT established relations with AMUT in Iran.

The relationship between MIT and Iran, particularly with AMUT, was shaped by the geopolitical climate of the Cold War and the Shah's regime. In the early 1950s, MIT began engaging with Iran in various technical and educational capacities. This collaboration led to the signing of agreements in 1974, focusing on nuclear engineering and technical education development. However, these agreements faced significant criticism, both from within MIT and the broader student and academic community. Concerns included potential contributions, nuclear proliferation, and ethical considerations about the authoritarian nature of the Shah's regime. Despite the controversy, MIT continued its engagement with Iran, involving faculty and influential alumni, and discussions of collaboration expanded into various fields, including oceanography and energy research. In the next chapter, we turn our attention to the legacy of Dr. Seyyed Nasr, who as the fourth chancellor of AMUT not only was instrumental in the formation of the MIT–AMUT partnership discussed here but also led other ambitious initiatives during his three-year tenure.

3

DR. SEYYED HOSSEIN NASR AT AMUT: BRIDGING SCIENCE, PHILOSOPHY, AND CONTROVERSY (1972–1975)

Sepehr Vakil: Was part of the vision that you had for the Aryamehr University, to create an engineering education so that people do engineering and they stay in Iran and that they work on behalf of Iran?

S.H. Nasr: Exactly. And they would be "Persianized."
—interview with Hossein Nasr, 2019

In this chapter, we examine the philosophy of Dr. Seyyed Hossein Nasr on science and technology and the significance of his tenure as chancellor of Aryamehr University of Technology (AMUT) from 1972–1975. A vast literature covers Nasr's ideas and intellectual contributions. Here we focus primarily on his thoughts regarding the philosophy of science and technology, examine his attempts to implement his ideas into practice while chancellor, and reflect on the significance of his legacy.

As we noted in the previous chapter, Nasr led the establishment of the formal relationship between AMUT and the Massachusetts Institute of Technology (MIT). He was also an essential force behind the Center for Humanities. His tenure as chancellor coincided with some of the most turbulent years in AMUT history, which included an attempt on his life by militant students. By all measures, Nasr was and remains an exceedingly complex figure—a Western-educated scientist and a renowned Muslim philosopher who continues to be known for his searing criticisms of

Western intellectual and scientific traditions. Notably, Nasr was a close ally and collaborator of the Shah's royal court and maintained a fraught relationship with leftist students both inside and outside of Iran. Throughout his tenure at AMUT, Nasr was in frequent communication with Queen Farah Pahlavi and the Shah himself, which separated him from activist students and estranged him from the broader political culture of the university.

Dr. Seyyed Hossein Nasr was the chancellor of the AMUT from 1972 until he resigned in 1975, following the deterioration of health and in response to being informed by SAVAK that he was on the terror list of the underground political party known as the People's Mojahedin Organization of Iran (PMOI; Nasr 2010). Unsurprisingly, Nasr remembers his tenure at AMUT as a highly stressful period of his life: "For three years, I was the Chancellor of Aryamehr university. That sounds like a short time, but it was like an eternity. Because the students were demonstrating every day."[1]

Nasr was the first Iranian undergraduate student admitted to MIT. He obtained a bachelor's degree in physics and then went on to obtain his PhD in the history of science from Harvard University in 1958 (interview with Hossein Nasr, 2019). Additionally, he earned a master's degree in geology. He felt strongly that the intellectual span and diversity of his credentials prepared him for the debates he encountered while at AMUT. In his interview with the philosopher and intellectual Ramin Jahanbegloo, Nasr said, "People could not say, 'You do not know what you are talking about, because you have a degree in physics and mathematics, but do not know the descriptive sciences'" (Nasr 2010, 45). Though he was offered a teaching position at Harvard, Nasr returned to Iran in the late 1950s and became a professor of Islamic sciences and philosophy at the University of Tehran. He served as the vice chancellor of the University of Tehran from 1968 to 1972, and during the time he was serving as the chancellor of AMUT from 1972 to 1975, he established the Imperial Iranian Academy of Philosophy in 1974 to promote and conduct comparative research and studies on Iranian and Islamic philosophical thought and traditions.

It is important to understand that Dr. Nasr comes from a wealthy and highly educated family, with deep roots in Islamic philosophy and Sufi teachings and practices. In our interview with him, he spoke proudly of his family origins and background:

My father was a doctor. A royal physician to the Qajar kings. Very famous doctor. My grandfather also was a royal physician. That's how we received our name 'Nasr'. Nasr was a title given from a Qajar king to my grandfather. My father, in addition to that, was one the greatest authorities of the Persian language and philosophy and literature and humanities in Iran. Was called 'ostad-e-ostadan' (the master of the masters). He was a very famous scholar, in addition to being a famous physician. So, I was born into this home. My mother came from a very religious family. Her grandfather was Sheikh Fazlollah Nuri.[2] (interview with Hossein Nasr, 2020)

While he earned multiple academic credentials from the best institutions available in the Western world, it is evident that his early education in Islamic philosophy and sciences had a profound impact on his intellectual and political views on science and technology.

To assess his philosophical position on science and technology, in addition to his published works (such as Nasr 1976, 1993, 2010), we rely on several key sources, including three interviews we conducted with Nasr in person in his Washington, DC, office on the seventh floor of the George Washington Library in 2019 and 2020. Further, we also draw heavily upon archival records pertaining to a lecture he gave at MIT in 1973, at the precise moment that he was fortifying the relations between the two institutions. Let's begin there.

NASR DELIVERS HISTORIC SPEECH TO MIT FACULTY AND STUDENTS DURING A 1973 VISIT

On a cold early spring day in 1973 at MIT, the leading scientific institute of the world, professors and students at the Technology and Culture seminar series gathered for an in-person address by Nasr. The backdrop and circumstances of the visit are remarkable. In his official capacity as chancellor of AMUT, Nasr's seminar was part of a multiday visit where he was presumably courting his alma mater as the intellectual figurehead and institutional leader of Iran's new university. A partnership with the world's leading scientific university was a highly sought-after objective, particularly for the Shah of Iran. However, as we recall from the previous chapter, MIT faculty and students, critical of the Shah's authoritarian politics, raised significant concerns and objections. Thus, the conditions of a partnership between the two institutions were deeply tenuous and fraught with uncertainty and discord. It is within these conditions that the attendees of the MIT event gathered in the seminar room for Nasr's speech, titled, *Contemporary Man: Between the Rim and the Axis*. How would Nasr, figurehead of AMUT, alumus of MIT, represent his philosophical commitments to science and technology? We highlight the significance of this 1973 speech delivered to the very center of Western scientific and technological power. Despite being in Massachusetts in his official capacity as chancellor of the Shah's new university, Nasr did not squander this rare opportunity to assail what he perceived as the moral impoverishment of Western thought. Early in his speech, he gave his unbridled assessment of the philosophical foundations of Western knowledge production:

The crass positivism of the past century which has caused philosophy as perennially understood to become confused with logical analysis, mental acrobatics or even mere information theory, and the classical fields of the humanities to become converted to quantified social sciences which make even the institutions of literature about the nature of man inaccessible to many students and seekers today.[3]

Extending this critique of positivism to a reckoning with the exaltation of scientific objectivity, Nasr used metaphysical references in his critical assessment of objectivity.

No doubt modern science possesses a limited form of "objectivity" in its study of the physical world, but even in this domain this "objectivity" is encompassed by the collective subjectivity of a particular humanity at a certain moment of its historical existence when the symbolist spirit has become atrophied and the gift of seeing the world of the spirit through and beyond the physical world has been nearly completely lost.[4]

Nasr conveys that what passes for scientific objectivity in the modern Western world would not bring humanity closer to the nature of existence, and to claim otherwise is an act of arrogance and ignorance. He went on to say, "It is as if an audience of deaf people testified together that they did not hear any music from musicians playing before them and considered the unanimity of their opinion as a proof of its objectivity."[5] Going further, he asserted that the modern theory of evolution is "merely a scientific illusion" and a consequence of modern "amnesia."

THE NEED FOR AN ISLAMIC AND SACRED SCIENCE

To escape this amnesia, Nasr argued AMUT engineers and scientists and students must have a much firmer grounding in the historical and social knowledge of Iran and its grand contributions to world civilization. However, for Nasr, this modern "amnesia" is not merely something intellectual in nature or due to a simple absence of historical knowledge but, rather, it is also metaphysical. This was the basis from which he advocated for an Islamic science. In this conception, Nasr is following an idea that Ziaddin Sardar had expressed previously in an essay titled, "Islamic Science: The Contemporary Debate," in which he said, "in this perspective, the Islamic science becomes the study of the nature of things in an ontological sense" (Harding 2011, 374).

Nasr outlined core aspects of his spiritual philosophy during his MIT speech, asking rhetorical questions to the American attendees: "So does the very fact of human existence imply the presence of the centre and the axis and hence an inevitable connection of men of all ages with 'Man' as such, with the anthropos?"[6] Answering his own question, Nasr inveighed further, outlining his vision for a sacred science: "I have been trying to talk over the years about the revival of the spiritual understanding of nature, and one of the key terms that I have to use is precisely the sacred" (Nasr 2010). He talked about the limitations of scientific knowledge and about the unique power of Islamic metaphysics. With his MIT audience, Nasr was explicit about how his religious and spiritual background informed his vision and leadership as AMUT chancellor. Decades later, sitting in his office in Washington, DC, he remarked:

I had a lot of contact with religious circles in Iran. That's very important in what happened in engineering at Aryamehr. And I met some of the greatest traditional teachers of Islamic thought and philosophy in Iran. Allameh Tabataba'i, Seyyed Muhammah Kazim Assar, and people like that. I studied with them for years and years. So, I've had a kind of Mullah training also, in the traditional Islamic science, in parallel with being a professor at Tehran university. (interview with Hossein Nasr, 2020)

Through years of studying with eminent scholars of Islamic thought and philosophy while simultaneously serving as a professor, Nasr developed a specific understanding that merges the inner spiritual world with the external scientific and academic world. This perspective is evident in his view on the nature of existence, where he articulated the interplay between the seen and unseen forces that shape our understanding of the universe:

The religious man sees God as the Inward; the profane man who has become completely oblivious to the world of the Spirit sees only the Outward, but precisely because of his ignorance of the center does not realize that even the Outward is a manifestation of the Center or of the Divine.[7]

We note, importantly, that his critiques of Western paradigms were not confined to the realm of spirituality or Islamic metaphysics. There is and was an important cultural aspect to Nasr's views. In the MIT speech, to a room filled with American students and professors, he proclaimed: "The medicine men of Africa have had a *deeper* [emphasis ours] insight into human nature than the modern behaviorists and their flock because the former have been concerned with the essential and the latter with accidentals."[8] The cultural or racial identity of these "modern behaviorists" remained unnamed, but the implications were clear. Perhaps the cultural element in Nasr's worldview is most visible in his own articulations related to his abundant pride in the contributions of Iranian Muslims:

So much in mathematics was developed by the Greeks and the Muslims. Trigonometry and at the beginning, Algebra . . . advanced math like Set theory and things like that, which are later developments, up to college, all mathematics taken in high school are developed, actually, in our own country. And we don't modernize it. We study trigonometry at Alborz college and says "Sine, Cosine." These are all European words that are translated from Arabic, "Jeib" and "Tejeib." Sine means "pocket" in Latin, like "sinus problem" and Jeib in Arabic means pocket. That's what it was, it was turned from Arabic and translated into Sine. (interview with Hossein Nasr, 2020)

NASR'S VIEWS ON TECHNOLOGY AND THE ENVIRONMENT AND THE ISFAHAN CAMPUS

Intertwined with the abovementioned dimensions of Nasr's philosophical approach to science and technology, his scholarly writings, as well as his speech at MIT, direct special attention to humanity's fraught relationship with the natural environment. In fact, Nasr's epistemological critiques of Western paradigms are directly related to his assessment of modern technology's blatant disregard for ecological systems and processes.

Interestingly, his critiques here echo recent debates regarding the relations between advanced technologies and the environment. As he himself pointed out,

At that time, the environmental crisis had not begun. Although I had already written the first book about the environmental crisis. And I had given a series of lectures at the university of Chicago in 1966. (interview with Nasr, 2020)

He noted that the escalating environmental crisis has led many in the modern world to critically reevaluate the Western conception of humanity that has dominated since the onset of modern civilization.

In order to gain dominion over the natural environment, he [modern humanity] has created a situation in which the very control of the environment is turning into its strangulation, bringing in its wake not only ecocide but also ultimately suicide. . . . The problem of the devastation brought upon the environment by technology, the ecological crisis and the like all issue from the malady of amnesia or forgetfulness from which modern man suffers.[9]

In his tenure as chancellor at AMUT, Nasr seized an opportunity to put his ideas regarding the environmental implications of technology into practice. As we described in earlier chapters, shortly after the founding of AMUT in Tehran, a secondary campus was built in Isfahan, Iran's third largest city. Today, the Isfahan campus of AMUT has developed to become an independent university known as Isfahan University of Technology (IUT); however, at the time of its construction in 1974, it was planned as a satellite campus of AMUT, with a shared administrative structure and leadership. The original goal was to erect a vertical tower as the primary building of the Isfahan campus. However, Nasr criticized the architectural design and instead suggested a horizontally constructed tower. Nasr argued that a high tower in the middle of the university would break with the Persian architectural tradition. He reasoned that since there is no scarcity of land at their disposal—with 24 million square meters available for the Isfahan campus—they didn't need to build a vertical tower, which would have required energy for elevators, air conditioning, and so on. Instead, the tower could be horizontal, which would be in better harmony with the surrounding environment and conserve energy. He also suggested using brick in the building's construction and tried to revive traditional brick-building techniques of artisans in Yazd, Qom, and Isfahan (Nasr 2010, 119).

To convince the Shah of the validity of his decision, Nasr said to him, "We could have the same concentration horizontally and save much energy and be more in harmony with the surrounding environment." In response, the Shah, "laughed and said, 'I never heard of a horizontal tower. All right, go change the plan and bring it to me'" (Nasr 2010, 119). Nasr did so, and the building was constructed based on this new approach.[10]

REACTIONS TO NASR AFTER HIS MIT VISIT

Archival records capture the severe criticism of Nasr's presentation shared by student attendees during the Q&A after his lecture. The criticism continued in a statement received by the Technology and Culture Seminar from some participants, leading to an exchange of letters among faculty, deans, and even the president of MIT. These letters included discussions by Nathan Sivin, historian of science and professor at MIT, and Gordon Brown about their reactions to the content of Nasr's speech, critically assessing whether the MIT–AMUT collaboration was wise given Nasr's ideological positions and the tenuous political conditions in Iran at the time.

Furthermore, anonymous attendees of the seminar sent in harsh rebukes of Nasr and his ideas, which, according to the transcript of the event circulated among the MIT faculty and president, originated from Iranian MIT graduate students.[11] These rebukes, which started during the Q&A after the presentation, were followed by a statement received by the Seminar that distinctly expresses a critical and political viewpoint opposing Nasr and his positions in the seminar. The statement mocks Nasr's "religious position" as "essentially unIslamic" due to its rejection of political accountability. This was in response to Nasr's repeated claims that he was maintaining a politically neutral position—despite his conspicuous proximity to the Shah's royal court. During the event, Nasr proclaimed, "I am here as a philosopher, not a politician." The anonymous statement pointed out the contradiction of a university leader claiming neutrality while students at the university were being targeted for their political activities: "Are there not university students in jail under false charges—their guilt never proven, their cases never tried?"[12] The anonymous statement proceeded passionately:

The public strongly reacted to this fragmented image of a man who, as vice-chancellor of the state-governed Arya-Mehr University, occupies a political position proper. He denied the political aspect of his office. . . . Can a man be a spiritual philosopher, a bigot, an elitist, a sufi, and a university chancellor all at the same time? And still retain his integrity? Dr. Nasr, a high-ranking academic personality, gave full credibility to this suffocating situation by endorsing an elitist ideology and preaching a 'spiritual' philosophy which is vulnerable to being used for fascist purposes because it refuses to accept political responsibility.[13]

While criticisms of Dr. Nasr were extensive and, in some cases, severe, the correspondence between Sivin, Gordon Brown, and President Wiesner reveal that the president continued his support of developing the MIT–AMUT affiliation. In a letter to Professor Sivin, a renowned historian of science, Wiesner wrote:

I have the same concerns about the political situation and its stability that you express plus my concerns about the nature of his rule. I long ago learned to keep to myself my revulsion regarding the repressive nature of Soviet society, in exchange for the opportunity to learn about it, to help end the cold war and to provide some access to the outside world for Soviet intellectuals.

Wiesner ended his letter, stating, "In spite of the present very discouraging situation in the Soviet Union, I think that the effort was worthwhile and I wonder if we don't have an equivalent opportunity to play a constructive or supportive role in Iran."[14] Furthermore, as quoted in a *New York Times* article covering the controversy, MIT's head of nuclear engineering, Kent Hansen, responded to the critics of the program: "Foreign students who come to MIT rise to become very important people in their society. . . . Since they do become influential, we feel that exposure to MIT is beneficial [15]" (quoted in Kifner 1975).

Support for the MIT–AMUT partnership from leaders like Wiesner and Hansen ensured the emerging partnership between the two nations would survive the harsh criticisms. Challenges to both the content of Nasr's philosophical positions on science and technology, or concerns for supporting an autocratic Iran, were ultimately no match to the steely-eyed Cold War political calculus of Soviet containment. Nasr's controversial views and critiques of Western paradigms notwithstanding, MIT's leadership ultimately endorsed the relationship with the Shah's new university. But how were Nasr's views received and understood by the Shah of Iran himself? How did Nasr's strong critiques of Western scientific paradigms square with the Shah whose reputation was almost synonymous with Westernization and Americanization?

NASR'S RELATIONSHIP WITH THE SHAH OF IRAN

Contrary to our own expectations, we learned in our research that the Shah requested Nasr leave the University of Tehran and come to AMUT precisely *because* he hoped Nasr would implement his philosophical ideas about science and technology there. The Shah was well aware of Nasr's writings, and his critical assessments of Western epistemological paradigms did not necessarily align with the Shah's modernity. And yet, given that Nasr was academically trained in both science and the philosophy of science, his critiques were treated seriously by the royal court of Iran. As Nasr himself recounted:

Shah told me, he said: "Dr. Nasr, I know what you've written, what you've said and reports about you. I know your educational background, you always said that we have to Persianize (Iranianize) what we receive from the west, make it part of our own culture and I have a lot of respect for that idea. Here it is, see what you can do with it. Aryamehr university is very close to my heart. It's Iran's best university, and I've chosen you to become its president." (interview with Nasr, 2020)

Further, Nasr's affinity for and knowledge of MIT was essential to the prominent position he held in the Shah's mind. According to Nasr, the Shah always introduced him by mentioning his personal ties to MIT, as opposed to Harvard University, where Nasr had obtained his PhD (Nasr 2010, 116). As Nasr clearly explained, "When the Shah would introduce me to a foreign head of state, he would say, 'Nasr is from MIT,' but would never say that I was also from Harvard, because MIT was sort of his ideal" (Nasr 2010, 116). In Nasr's conversations with Queen Farah, he recalled that AMUT was referred to as "the Shah's university," which indicates the essential role it played for his modernization projects. In the interview with Nasr, we asked him: "So, did the Shah have any problem with your idea of bringing in Iranian and Islamic history and culture?" Nasr replied:

No, just the other way round. . . . He said, we have to learn Western science and Western technology. We have to integrate it into our own culture and see . . . he was always painted as being a modernist. In a sense he was a modernist, but he was also very much interested in his own country. He wanted Iran to modernize. Not that the modern world to just take over Iran and change it from what it was. (interview with Nasr, 2020)

At the core of the interactions between the Shah and Nasr, we observe a dilemma that both Iranian intellectuals and the state shared: how to adopt advanced Western science and technology while still preserving Islamic and Iranian cultural heritage and values?

In 1968, Nasr attended the well-known conference on education in Ramsar, a city in the north of Iran, which is where the Shah first learned about Nasr's opinions and criticisms on these matters. As Nasr recalls, in his interview with the Foundation for Iranian Studies, he inveighed,

the institutions of universities and its management and organization can not be separated from the culture of its hosting nation. We can not develop our universities in a vacuum free of our cultural heritage. (Hossein Nasr, interview with Foundation for Iranian Studies, 1982–1983 [originally in Persian])

Immediately after his presentation at the Ramsar conference, the prime minister Amir-Abbas Hoveyda told him that the Shah had chosen him because he was the only one who offered a serious critique of the challenges of modernization, and had even circled his name with red ink, exclaiming, "Nasr must certainly become the chancellor of one of our universities" (Hossein Nasr, interview with Foundation for Iranian Studies, 1982–1983 [originally in Persian]).

To be sure, Nasr was not against the adoption of Western technology and scientific knowledge. Rather, what he vehemently objected to was "just blindly copying Western technology" (Nasr 2010, 116). In the opinion of the Shah and the Iranian royal court, Nasr was an ideal chancellor for AMUT because of his familiarity with MIT, his own scientific knowledge, and his embrace of the importance of technological development, as well as his concerns regarding how to adopt these advances in accordance with the cultural and religious context specific to Iran. As several scholars have noted, the Iranian state in the 1970s rallied behind the concept of "returning to the self" in which "the self" was synonymous with Iranian cultural heritage.[15] Nasr said this of the Shah:

[He] knew that I was well acquainted with Western science. I also knew what technology was, but at the same time was a great champion of Persian culture and a critic of any form of science or technology that would simply decimate

3.1 Hossein Nasr (left), Firooz Partovi (right). Photo credit: Personal collection of Firooz Partovi.

our traditional culture. He knew that fact very well, and he also knew that I had written about the environmental crisis, against what we were doing to the natural environment as a result of modern technology. (Nasr 2010, 116)

In practice, how successful was Nasr in implementing his ideas at AMUT? Who emerged as his supporters, and who were the critics?

THE CONTESTED LEGACY OF THE CENTER FOR HUMANITIES

In addition to his leading role in facilitating the MIT–AMUT partnership, the establishment of the Center for Humanities at AMUT is Nasr's most significant contribution. The enduring imprint of the humanities as part of the very essence of AMUT is unmistakably a consequence of Nasr's ideological commitments combined with his administrative acumen during his time as chancellor.

In our interview with him in his Washington, DC, office, Nasr described extensively his vision for the Center for Humanities, as well as the challenges he encountered in developing it at AMUT. Nasr's ideas about the role of humanities in an engineering education date back to his undergraduate studies in physics at MIT. "We had to take one course in the humanities every term, but I took a lot more because I was so interested in the various branches of this field" (Nasr 2010, 39–40). Simultaneously, he also took courses in humanities at Harvard University related to Western and Oriental art. Nasr's initial forays into the humanities ultimately led him to pursue philosophy for his doctoral work, rather than continuing in physics. However, as historians Stuart Leslie and Robert Kargon have explained, Nasr's engagement with the humanities at AMUT was motivated by a different purpose than the presence of humanities at MIT at the time of Nasr's studies. They wrote:

Nasr interpreted his charge at AMUT as proving to the Shah that the university could train engineers who could compete on a world level without abandoning their cultural values. He had been a student at MIT during the years when strengthening the humanities and social sciences had first become a priority and drew a completely different lesson than had Brown and his colleagues. MIT administrators considered the humanities a matter of broadening the horizons of future engineering leaders and corporate managers. Nasr believed that in the

Iranian context the humanities were a question of national identity and purpose, the bedrock of a technical education, not a cultural veneer. (Leslie and Kargon 2006)

In addition to national identity, as we have noted, Nasr believed that to produce ethical scientists and engineers who were responsive to cultural, environmental, and spiritual matters, engineering students needed a specific technical education rooted in the social and cultural history of the nation. Nasr believed firmly that conventional Western ways of teaching science and technology to STEM students would estrange Iranian students from the essence of their history and culture. He feared a generation of Iranian scientists and engineers who would be wholly incapable of addressing the real needs and actual concerns of their own country.

As Nasr described, his agenda pertaining to elevating the humanities at AMUT faced great opposition from his colleagues: "When I went there I knew that it was going to be difficult to change the view of Phds from MIT, Stanford, Berkeley, Imperial College of London. . . . I was not going to be able to change overnight the views of these people concerning science and technology . . . many of whom were leftists" (Nasr 2010, 117). When asked to explain the source of his colleagues' objections to expanding the role of humanities, and how he ultimately overcame them, Nasr stated:

I had a good argument. I said: "Look I'm a graduate of MIT. In 1950 I entered MIT, Killian was then president of MIT, and just in that year, he begun to build a beautiful building for the Humanities—or just before I finished that year—and, have humanities in university. Which at first students of science and engineering took like ping pong. You know, two hours of history is like playing ping pong. They didn't take it seriously, but now, even then, few years has passed, and MIT is already known in the fields of social sciences and political sciences. It is a very important part of the curriculum and being followed by other major technical Universities like Caltech and places like that, RPI (Rensselaer Polytechnic Institute). Anyway. I was able to convince them. (interview with Nasr, 2020)

Nasr understood his Western credentials in STEM as a key advantage that was essential to his ultimate success in persuading the AMUT faculty: "It was very difficult, but anyway I was strong enough, and because I came from a scientific background, I could withstand their attacks, to accept that all the students at the university take some courses in Humanities" (interview with Nasr, 2020).

Nasr asked Morteza Anvari, the first chair of the Mathematics Department at AMUT and the founder of the Computer Science Department, to lead the Center for Humanities. Nasr explained, "Immediately upon becoming president, I was able to establish a department of humanities. I brought Haddad Adel[16] and William Chittick,[17] who were among the most brilliant of my students, along with several other gifted faculty to Aryamehr University to start this program" (Nasr 2010, 117). This is how Nasr "brought the whole dimension of the humanities to the university" (Nasr 2010, 117).

Anvari invited prominent artists to teach at the center, such as well-known modernist and leftist poet Ahmad Shamloo, prominent musician Hormoz Farhat, and great painters such as Hannibal Alkhas and Iran Darroudi. For Nasr, the Center for Humanities would emphasize contributions Iran's historical contributions to scientific and technological innovation. We quote him here at length:

One aspect is just bringing the history of the development of science as part of the education of engineers or scientists. And secondly, what I tried very hard, was, I said, look there's a great deal of engineering which exists in our civilization before the coming of modern engineering. Take the system of the marvels of irrigation engineering, which underground they're able to dig in such a way that water on the mountain comes down on the valley. None of you now could do it. I said look, is there anyone in this room? You all have PhDs in mathematics, and structural engineering and so forth. Can you design the dome of Taj Al Molk in Jameh Mosque in Isfahan[18] or not? So, I tried to show them that is very, very important for us to also in engineering, revive teaching students all the remarkable things that we did in engineering. (interview with Nasr, 2020)

Earlier we alluded to faculty resistance to the Center specifically and more broadly to Nasr's vision for expanding the role of humanities into what might be considered a classical engineering education. Notably, our research revealed analogous skepticism from AMUT students. First, it's important to consider that once the center was established, not all departments required students to take classes from the Center. For instance, the Electrical Engineering department did not require students to take any courses from the Center, while math and computer science required students to take at least ten units.[19] Thus, students had varying levels of exposure to the cultural, humanistic, and religious offerings associated with Nasr's Center. All of the former AMUT students we interviewed, however, conveyed they had taken at least one if not several classes at the Center during their student years. In asking for their reflections, one of the most surprising findings was recollections of student concerns regarding the hidden political agenda attached to the Center.

One of our interviewees explicitly stated the Center was established to depoliticize the student body at AMUT, and several more referenced how the Center *superficially* allowed the students to have democratic access to thinkers and artists who were critical of the state. They lamented that the university hired leftist intellectuals but that it had done very little to genuinely broaden their intellectual horizons. As one of our interviewees, recounted:

I remember once Shamloo was reading one of his banned poems in a class, when they called the person in charge for organizing these (extracurricular) classes and asked him to let Shamloo know not to do this again, otherwise . . . It wasn't like they were paying them much . . . it was 5000 Tomans a semester. I remember that the poor guy used to come to the university once a week for four months to get 5000 Tomans. His course was called "experimental studies of Persian" . . . but that was the only thing he wouldn't teach.

Another interviewee recounted a personal experience that reveals the complex and, at times, ambiguous political orientation students ascribed to the Center. One student shared that the Center appeared to invite political discourse, but not without risks:

The only potential place for political conversations at the University was the humanities center, established by Nasr. Even then, it was very limited as the students used to avoid having those conversations. As I told you William Chittick was teaching Islamic mysticism at the university. I remember once, he was talking, and I raised my hand to debate with him. . . . When I finished, he denigrated my comments as having a Marxist point of view. I stopped going to his class for the next few sessions so he would forget about me and that conversation. This is how the atmosphere was. We were worried to be labeled as Marxists. Although almost everyone was sympathetic to the left, no one wanted to be recognised as one. Especially in a public place like the classroom.

While depoliticization of the campus does seem to have been a goal Nasr shared with the state, our research revealed no clear indication that Anvari invited Shamloo and other critical thinkers to the campus specifically to depoliticize the students. In contrast, Dr. Nasr believed that what he assessed as the simplistic worldview and historical ignorance of STEM students at AMUT played a role in their radicalization, and that a well-grounded and humanistic education in arts, history, philosophy, ethics, literature, spirituality, and the cultural heritage of Iran could serve as a remedy for their political misconceptions and misgivings. In his own words:

The reason the leftists could trick these people [professors and students] at AMUT was the limitation of their knowledge as well as them being very young and inexperienced. It is not irrational that communists and leftists chose engineering departments, instead of law and literature departments, for recruiting allies. Their scientism combines with their lack of familiarity with their society played an essential role. (Hossein Nasr, interview with Foundation for Iranian Studies, 1982–1983 [originally in Persian])

Remarkably, Nasr effectively mobilizes his philosophical critiques of Western scientific paradigms toward an explanation for engineering students' politicization or, in his assessment, their gullibility. We return more substantively to the question of students' political identities in the final chapter, but we note now that our evidence does not support the theory that student's supposed knowledge deficit was the cause of their politicization.

In an obituary written for Anvari, Shiva Farahmand-Rad, an AMUT alumni, said that after Anvari left his position and was replaced by Hadad Adel, the Center developed even more Islamist tendencies. Interestingly, Hadad Adel had also originally been a physics student during his undergraduate studies, but after he listened to Nasr's presentation on the Muslim philosopher Mulla Sadra, he decided to study philosophy. Notably, Hadad Adel has also held prominent positions in postrevolution Iran. Fereydoon Hojabri, a former professor and deputy chancellor at AMUT, similarly describes the increased Islamicization of the university in his memoir of his days at AMUT (Hojabri 2016). Recent scholarship has argued that this era in AMUT's history was a precursor to the Islamization of Iran's higher education system in the postrevolutionary era (Pourjavady 2013).

Given the anti-imperialistic, revolutionary culture prevalent at AMUT, one might expect that Nasr's philosophical critiques of Western scientific paradigms would resonate with leftist students who were fundamentally protesting the US-backed autocratic regime. After all, if there is a commonality to be drawn between leftist AMUT students and Dr. Nasr, it in their shared critiques of the West. For Nasr, this specifically entailed critical assessments of Western knowledge production and epistemic regimes of rationality and scientific objectivity. For the students, they were anti-West primarily in their opposition to the Shah and his policies. As it pertains to Nasr specifically, his epistemological critiques of Western knowledge were ultimately overshadowed by his proximity to the Shah. As one of our interviewees stated, "There definitely was an alignment of Nasr's mindset and the type of political culture that the Shah wanted to run in the society." He continued by making a severely critical assessment of Nasr as chancellor and, more broadly, of his legacy:

So, in my opinion, Nasr is a regressive intellectual. In Iran, Nasr and his community have always been on the side lines. He didn't have much impact in the intellectual world of Iran. His main impact was due to his relationship with the palace or institutions abroad. I must add that he has valuable work. I've read his books. He has good analysis when he talks about old Islamic sciences. Or introduction of Islam Sufism or mysticism. But his views and analysis on the modern world, in my opinion, are regressive. (interview)

In our interview with Professor Ahmadiyan (name pseudonymized) who was teaching at AMUT at the time, he also echoed this sentiment:

This institute (Center for Humanities) is still running with some respectable individuals working at it, but philosophy was never taken seriously in Iran. As it got mixed with politics. Science and technology also got mixed with politics. In fact, the Shah and the Islamic Republic regime wanted this to happen, but politics never overtook science. There are aspects to science that make it more powerful than politics. It wasn't possible for religion to win over science and technology, the way Nasr wanted. (interview with Ahmadiyan, 2022)

NASR'S CONFRONTATION WITH STUDENT AND FACULTY ACTIVISTS AT AMUT

Nasr's chancellorship lasted only three years. When he resigned his position in 1975, he was dealing with heart issues that developed as a result of the stress of running the university during the political upheavals of the 1970s (Nasr 2010, 116). Nasr was always in close communication with the royal court of Iran. However, he did not perceive this proximity to them as a reflection of his political position. Instead, he considered himself an academic who was disinterested in politics, someone who straightforwardly shared his thoughts with the important men of the state when called to do so. Nevertheless, to the students of AMUT, he was perceived as the most powerful political figure at the university—and as a person whose politics were closely aligned with those of the Shah and his hand-picked allies within the regime. This perception manifested dramatically in the death threats and the assassination attempt that nearly ended his life.

Importantly, student and faculty reactions to Nasr were not unprovoked. During his term as chancellor, harsh reprisals were made against any students who went on strike. Nasr managed to do this by convincing the council of professors to agree to act against the strikes to maintain the high educational quality intact (Hossein Nasr, interview with Foundation for Iranian Studies, 1982–1983 [originally in Persian]). Nasr's explanation for such stern measures was that he believed the strikes could cause the serious deterioration of the quality of education at the school.[20] As he explains in an interview with Jahanbogloo, upon accepting the position of AMUT chancellorship, he had informed both professors and students that he "will not sacrifice the quality of education here [at AMUT] for anything in the world" (Nasr 2010, 121). As a result, students who defied his authority were barred from campus.

This action, along with the University Guard's crackdown on student demonstrations, resulted in the eventual assassination of the head of the AMUT university guards Sarvan Yadollah Nowruzi by OIPFG and the detonation of a bomb in Nasr's office.[21] Nasr survived this attempt on his life only because he fortuitously happened to arrive at work that day thirty minutes later than usual (Nasr 2010, 122). Four and half decades later, when reflecting on this episode and the circumstances that brought his term as chancellor to an end, Nasr stated:

I had some heart problem[s], I had to resign as president of the university. There was always rabble-rousing. They wanted to assassinate me; they killed my guard! You cannot imagine what was happening. Mudjahedin—these communists—these people were very murderous. (interview with Nasr, 2019)

Before Nasr's appointment, Reza Amin, the outgoing chancellor, told Nasr that AMUT was the toughest university to manage due to its politicized professors and students and, simultaneously, the quality of its professors and students (Hossein Nasr, interview with Foundation for Iranian Studies, 1982–1983 [originally in Persian]). Nasr himself has discussed the roots of political difficulties at AMUT. In addition to their alleged naivete, he noted that during his presidency, almost all AMUT professors were younger than thirty years old. Furthermore, he also attributed their politics to the material conditions many faced in their home villages. He said that most professors and students were from lower socioeconomic backgrounds and were deeply pained by the deep economic discrepancies they observed between their home villages and affluent parts of Tehran (Hossein Nasr, interview with Foundation for Iranian Studies, 1982–1983 [originally in Persian]).

AMUT had two dilemmas or *gereftari* [emphasis ours], one was all the professors were so young . . . most of them except for one or two . . . were between 25 and 30 years old. It was mind-blowing how young they were. Second, almost 80% of them had come from impoverished strata of the society . . . they were different from middle class in Iran, or for instance, Tehran University's professors, they seemed like villagers, it was overly visible and painful for them. (Hossein Nasr, interview with Foundation for Iranian Studies, 1982–1983 [originally in Persian])

As we've described, Nasr offered several theories to explain AMUT students' particular affinity toward political activism. Drawing on his experiences across multiple institutions in Iran, a final explanation Nasr provided for the politicization of AMUT students was their especially strong connection to political activists outside Iran relative to other universities in the nation (Hossein Nasr, interview with Foundation for Iranian Studies, 1982–1983 [originally in Persian]). Indeed, since the professors of AMUT were quite young—many having experienced student activism in Europe and North America in the late 1960s—the campus was influenced by political parties and movements both outside the country and by the confederation of Iranian students in Europe and the US. This led to further politicization compared to other campuses (Hojabri 2016). As we discuss in depth in the following chapter, many students were arrested, some were imprisoned and tortured, and some were even killed.

As we have demonstrated, Nasr's relationship to leftist students in Iran was complex and contested. He held multiple theories to explain their activist tendencies, many of them unfavorable to the students and their families. Yet in our interview with him, he also conveyed a mournful sense of sympathy and solidarity with the students. He described several incidents where he asked the authorities to release students who had been arrested. However, in one case, the student detonated a bomb after his release. Nasr described in an interview with Jahanbegloo his unsuccessful effort to intervene prior to the incident:

I said to him, "I am putting your hand in your father's hand; for their sake, go back to your studies. Do not do these things. You can reach your goals of making the country a better place if you are a person who is educated, who has a degree." Anyway, . . . A few days later a bomb was exploded in Cinema Capri in Tehran that killed fifteen, twenty people, and this boy had done it. (Nasr 2010, 125)

Despite these sympathetic statements, Nasr himself was keenly aware that the policies he implemented at the time of his chancellorship have been interpreted as favoring the Shah over the student activists and placing the interests of the Shah above the interests of the nation as a whole. However, Nasr disagreed:

Some people may criticize me, but there was no possibility of running that university or any other university without keeping order. And if you did shut down the universities for several years in order to create a new, peaceful ambience, there would have been no higher education in Iran, and Iranian families, especially the families of the poorer students, would have been even more unhappy. (Nasr 2010, 125)

Ultimately, Nasr perceived the political activism of AMUT students as standing against their own personal academic development as well as against the economic development and technological advancement of Iran. According to Nasr, the political activism of the students was an obstacle to modernization and technological advancement. Furthermore, for Nasr, this stance did not contradict his philosophical critiques of Western scientific paradigms.

In this chapter, we have discussed in depth Nasr's legacy at AMUT. Given the complexities and contestations we've examined, how might we best conceptualize his enduring significance for Iranian history, for contemporary discussions on the politics and ethics of STEM education, and as a global figure? By the time he had become chancellor at AMUT, Nasr had emerged as a global figure. That being the case, it was not only the leftist students and faculty at AMUT whom he had either infuriated or impressed. He was a provocative Islamic intellectual in a highly political position during a turbulent and fraught political moment in Iran's history. In the US, Nasr's politics had sparked animated debates and discussions about both the Iranian diaspora as well as among academic leaders in influential places such as MIT.

Nasr's global significance is evident in the intrigue he garnered not just within academic and scholarly circles but also in the political domain. For instance, according to a declassified document from November 1976, the US ambassador in Tehran had advised Nasr to get in touch with Henry Kissinger, who was then US secretary of state. It was suggested that Kissinger would be interested in hearing Nasr's perspective on the escalating tensions between the Shah and Reza Baraheni, a widely regarded public intellectual and leading critic of the Shah. The ambassador's remarks were as follows: "It is obvious that the GOI [Government of Iran] has become concerned about what it regards as the lies which Baraheni is telling about it" (letter from the US Embassy in Tehran to the secretary of state, November 1976). Kissinger's inquisitiveness regarding Nasr's perspective emphasizes the substantial impact and discussions influenced by Nasr's global prominence. Furthermore, Nasr being dispatched by the Iranian government to suppress the opinions of an Iranian scholar in the US underscores the nuanced nature of his legacy concerning freedom of expression in the years leading up to the revolution.

Furthermore, while in contemporary discussions Nasr is largely viewed as a scholar and commentator on Islamic philosophy, he should also be recognized as a serious global voice that can inform ongoing work in ethics, policy, and STEM education. For instance, it is remarkable to consider Nasr's ideas in the early 1970s regarding the culpability of Western paradigms to ecological crises. We contend that despite the political tensions associated with his proximity to the Shah, Nasr should be properly viewed as a prominent Iranian Muslim intellectual who made profound contributions to global conversations about science, technology, environmentalism, and society. He did this both as an intellectual as well as one of the architects and leaders of a regionally significant engineering institution.

4

THE RISE OF FACULTY AND STUDENT ACTIVISM AT AMUT

From its founding in 1966 until the revolution of 1979, despite its formal relationship to the royal court of the Shah, Aryamehr University of Technology (AMUT) became a highly politicized educational environment. Students and faculty at AMUT, similar to at other leading institutions of higher education in Iran, had wide-ranging demands regarding numerous social, economic, and political issues. Their politics and demands became radicalized in the early 1970s, and eventually came to include freedom of assembly, academic freedom, freedom of expression, and the overthrow of the Shah and his monarchical regime. During this decade, several AMUT students, professors and alumni even played leading theoretical and organizational roles in the Marxist and Islamist armed struggle movements against the Shah (Haqshinas 2020; Ebrahimzadeh 2021). AMUT students and faculty enthusiastically embraced the growing anti-Shah movement and also actively contributed to its intensification by participating in the wider protest movement, ultimately leading to the 1979 revolution that dramatically altered the geopolitics of Iran and the broader Middle East.

Despite a growing body of interdisciplinary research on student activism in diverse global contexts (e.g., Heffernan and Nieftgodien 2016; Hodgkinson and Melchiorre 2019; Pensado 2013), relatively few empirical and archival studies on student movements in Iran have been performed.

This stands in contrast to the well-documented activism of Iranian student groups in the US and Europe, their transnational networks, and global solidarities against US Imperialism (Matin-asgari 2002; Nasrabadi 2014; Aquilina 2011; Shawkat 2012; Moradian 2022). However, we must qualify this discrepancy in two ways. First, an abundance of memoirs, pamphlets, oral histories, and Persian-language monographs have explored various facets of the revolutionary student movement. We draw upon these throughout this chapter. Second, we also would be remiss if we ignored the significant contributions by Iranian diasporic scholars who may have not examined engineering student activism specifically but have produced important work in adjacent fields of study. For example, Shahrzad Mojab (1991) is a notable example of a scholar of Iran's educational system who has conducted pioneering research into the general history of faculty activism at AMUT during 1977–1979, focusing on the faculty's struggle for autonomy from the royal court and Iranian state. Other scholars have delved into the history of militant and political organizations of the time and the broader militant movement (Vahabzadeh 2010; Rahnema 2021; Abrahamian 1989; Behrooz 2000; Elling 2021). The efforts of these scholars have directed our focus toward primary sources and aid our comprehension of the broader local and global resistance landscapes. Moreover, they have shed light on the national transcampus and transnational connections and sources of intellectual and organizational influence, which have been pivotal in shaping and fostering political identity and aspirations at AMUT. Our research contributes to this important body of work through a focus on both student and faculty activism at AMUT.

As we have described, AMUT was the most prestigious engineering university in Iran and simultaneously one of the most significant sites of protest against the Pahlavi regime (1925–1979). Other universities, including the University of Tehran and the Polytechnic University in the bustling capital, Tehran, as well as Azarabadegan University in Tabriz, were also hotbeds of political activity. The activism across these campuses was interconnected, and any account of activism at AMUT must include references to these other universities, particularly those in Tehran. Moreover, the protests at AMUT and other campuses in Iran rest on a legacy of student protest in Iran.

A BRIEF HISTORY OF UNIVERSITY STUDENT ACTIVISM IN IRAN

The establishment of AMUT was concurrent with a specific period of student activism in Iran, namely, the rise of militant resistance to the US-backed Shah of Iran. It should be noted that not all students at AMUT were involved in student activism, and only a minority were members of established militant organizations. Nevertheless, the political atmosphere on the AMUT campus was intertwined with the broader trends of militant activism across the nation. Of course, the history of student activism in Iran predates the establishment of AMUT.

Student activism in Iran is commonly traced back to the 1940s. Following the occupation of Tehran by the Allied forces in 1941, the authoritarian rule of the first Pahlavi monarch came to an end. During the period between 1941 and the 1953 coup, the regime of the second Pahlavi leader was comparatively weak, leading to a decade of relatively unrestricted political expression in the absence of a strong and repressive central government. The student movement emerged in the mid-1940s, and several student organizations and groups were formed at major universities, particularly at the University of Tehran (Yazdi 2004; Ziyazarifi 2016).

The desire for social and political freedom continued to intensify among Iranian students, with calls for university autonomy becoming an increasingly central demand. According to Mojab (1991), autonomy was interpreted as financial independence from the Ministry of Finance and administrative independence from the Ministry of Education. The infamous failed assassination attempt on Mohammad Reza Pahlavi, the second Pahlavi shah, at the University of Tehran campus in February 1949, led to renewed efforts by the state to militarize the campus and discourage political activism among students (Ziyazarifi 2016; Amirkhosravi 2015).

Despite the active role played by student activists in the movement to nationalize the Iranian petroleum industry under the leadership of Prime Minister Mosaddegh, the CIA- and MI6-sponsored coup of 1953 led to the downfall of Mosaddegh and the installation of a new government that brutally suppressed the nationalization movement, and also suppressed political organizing by Marxists as well as by the pro–National Front.[1] Universities were not spared. The state implemented several harsh measures to subdue and challenge the autonomy of the University of Tehran, including interfering in the election of its chancellor and taking responsibility for educational planning. As a result, the power of the university to make its own policies became increasingly restricted, and its authority to hire faculty and staff was significantly diminished. Consequently, the autonomous status of the University of Tehran was abolished (Mojab, 1991).

Less than two months after the coup of 1953, on December 7, students were protesting the visit by Richard Nixon (at the time US vice president) to Tehran. Police forces entered the University of Tehran's campus, and three protesting students were killed. Even today, December 7 is still known as Student Day (*Rūz-i dānishjū*), to commemorate the memory of the three students who were killed at University of Tehran (Ganjavi and Mojab 2018).

Over the next decade, the postcoup regime intensified its suppression of students and various nationalist and leftist political organizations. Having consolidated their political position, in the late 1950s, the Shah and his regime relaxed some of their most repressive control tactics. This led to a slightly freer and somewhat more open political sphere than had prevailed even a few years earlier. This, in turn, led to a resurgence, though limited in scope, of student activism. However, even this (relative) political openness was short-lived. The students were soon disillusioned with the suppression by the regime at the University of Tehran in January 1962 (Jazani n.d.), and by the later even more brutal suppression of the June 1963 uprising (Haqshenas 2020).

This prolonged period of political repression led to the emergence of underground political leadership and fueled the radicalization of activism in the late 1960s. By the beginning of the 1970s, the reformist and nationalist groups had been eclipsed by militant Marxist and Islamist organizations that were recruiting heavily on university campuses (Abrahamian 1982). The intelligence service of the Shah's government kept a close watch on the university, monitoring the behavior of students and professors daily (NLI, Document Number 230–035171). It was during this period that AMUT was founded.

THE POLITICAL CULTURE OF AMUT

In a dramatic autobiographical novel, published while the author was in exile in 2018, Shiva Farahmand Rad vividly portrays his harrowing journey as a former student activist at AMUT. His gripping autobiographical narrative reveals the political and cultural atmosphere at the university, as well as the broader social history of Iran in the tumultuous 1970s and beyond (Farahmand Rad 2018). The book begins with his confession of love to Azadeh, a former classmate, made decades after the events that irrevocably altered the course of Farahmand Rad's life. In his second year at AMUT, Farahmand Rad was transformed from a starry-eyed romantic to a fearless activist, who was subjected to interrogation, trauma, and, ultimately, a life in exile.

As the story unfolds, Farahmand Rad paints a seemingly idyllic picture of a sunny spring day in 1975, when he was happily heading home on a street near AMUT. However, the tranquil atmosphere was abruptly shattered when SAVAK agents, (who had been monitoring his political activities) arrested him. They forcibly took him back to his apartment for a search, even going so far as to play every audio tape they found to check for incriminating content. Among these tapes was a hard-won recording of a Turkish song by Soviet Azerbaijani singer Rashid Behbuduv. Terror gripped Rad's heart as he frantically thought to himself, "What if the SAVAK agent realizes the singer is a Soviet citizen and that this music is forbidden in Iran?" Suddenly, the voice of one of the agents interrupted his thoughts: "Found it!" The agent had stumbled upon a container of iron perchlorate, salts derived from perchloric acid, which is used for manufacturing batteries, tucked away in his closet. The agent barked, "What is this? Do you use it to make bombs?" Farahmand Rad stammered in response. "No, it's just a hobby I have. I use it to repair electronic devices, like radios and televisions."

The interrogation was far from over. Another agent suddenly declared, "Found it!" This time, the discovery was a collection of books, including works written by Samad Behrangi (1939–1967) and Gholam-Hossein Sa'edi (1936–1985), both leftist fiction writers and social critics whose works were either banned, censored, or deemed "politically deviant" by the state.[2]

Farahmand Rad's attempt to defend his possession of these banned books was abruptly cut short by the third shout of "Found it!" coming from another of the SAVAK agents. This time, the agent held up a poorly copied paper. Farahmand Rad felt a sudden pang of fear, realizing that there could be more to this search than he initially thought. These thoughts went through his mind:

What if my roommate is hiding something? What if my roommate is a member of a hidden political organization? He has mentioned a friend of his, with the name of *Aqa*,[3] a few times. Who could *Aqa* be?

Farahmand Rad let out a sigh of relief as the SAVAK agents revealed that what they had found was merely a copy of a page from a textbook he was studying at AMUT. The search and interrogation finally concluded, but before departing, the agents warned him by saying, "This time passed safely, but we are watching you." The experience left Farahmand Rad shaken to the core—both mentally and physically (Farahmand Rad 2018, 4–10).

The account of Farahmand Rad resonates profoundly with the stories shared with us in oral histories we have conducted by dozens of Sharif alumni. These types of incidents were common occurrences at the university throughout the 1970s. The themes of secrecy and the threat of surveillance and detainment by SAVAK profoundly shaped the student and faculty culture at AMUT. The pervasive sense of secrecy among activists was matched by a constant threat of SAVAK's presence.[4] Several of our interviewees shared chilling memories of classmates and friends who were killed during guerrilla missions or executed after the revolution. One such informant, Hengameh, told us:

I remember in one of those strikes in 1970, students set a bus on fire. It was in response to the increase in price of the bus tickets and the fact that a student was killed—although SAVAK later announced that he had died of Septicemia. They had sent some security guards to the university—riot police—which were different to the ones you have these days—they were city guards. (interview with Hengameh, 2021)

This harsh reality of student activism and its severe consequences are echoed by another story that highlights the personal losses suffered by the academic community. Providing a poignant reminder of the dangerous environment these students navigated, another interviewee shared her own painful memory:

Another peer of our time was such a talented person coming from a small town from Kerman. In the second or third year, we noticed he is not coming to the campus, after a while we heard he was also killed. (interview with Maryam, 2021)

She explains the culture of secrecy as follows:

Nobody wanted to make their political orientation public, we would not know anything about their orientation before these people going underground [i.e, starting a hidden life as part of joining a militant organization], the political suffocation at the time is not comparable to what we are experiencing these days, these days everyone at least talks, but at the time you would not even dare to reveal your criticism to yourself. When someone was killed, others would not do something [i.e. mourn, or protest] which could make their orientation clear. (interview with Maryam, 2021)[5]

This atmosphere of fear and secrecy deeply influenced the way students navigated their daily lives and political expressions on campus. What fueled a culture of secrecy and suspicion that permeated student life at AMUT was partially the perception of varying levels of commitment and affiliation to organized leftist groups and ambivalence regarding political affiliations of students and faculty, which raises the question: How widespread was activism and activist affiliations among the student body?

CHARACTERIZING THE EXTENT OF ACTIVISM AT AMUT

Accurately representing the proportion of engineering students engaged in political activism was a significant challenge for several reasons. First, as was discussed in the section concerning the political climate of Iran in the postcoup era, widespread political suppression drove much of the activism underground. Second, as we will discuss in this chapter, the operational methods of militant organizations at the time were characterized by stringent confidentiality policies. This meant that even members within a particular organization or student activist cell were often unaware of each other's identities.

Additionally, it is essential to recognize that the number of politically active students varied over time, influenced by shifts in social movements as well as tactics of state repression. While activism experienced a notable upsurge between 1971 and 1974, the subsequent two years saw a decline in student engagement on campus. This downturn was attributed to SAVAK's success in neutralizing the leadership of certain organizations and internal divisions within others. However, beginning in 1978 until the onset of the 1979 revolution, a resurgence occurred in active student participation. The months leading up to the revolution saw a marked increase in politically involved students, spurred by the momentum of the burgeoning revolutionary movement underway in Iran.

Despite these constraints, we compared the total number of student admissions during these years with the total number of student casualties explicitly attributed to political engagement. As previously noted, according to an extensive list produced by Shiva Farahmand Rad, 135 students at AMUT were killed either before or shortly after the revolution due to their political activism. During the first ten years, there were 4,447 admissions. While exact figures for the final two years before the revolution are unavailable, it is reasonable to estimate approximately 1,000 additional students based on the trend of student admissions, bringing the total to around 5,447. Thus, 2 to 3 percent of students attending AMUT during those years were killed due to their activism. It's important to note that this figure strictly represents militant student activists who were not only officially affiliated with a political organization but also engaged in direct physical confrontations with the state. Notably, this figure does not include students who were proximal and sympathetic to student activists but not directly affiliated with particular organizations, which, based on our research, comprised the vast majority of engineering students at AMUT. We turn now to examine the specific ideologies that student activists and their allies embraced and how these ideologies opposing the state took hold despite significant constraints on academic and intellectual freedom in Iran.

POLITICAL ACTIVITIES AND DEMANDS AT AMUT UP TO THE SIAHKAL INCIDENT

In 1970–1971, various student groups were already active at the AMUT campus. For example, the Islamic Student Association (anjuman 'islāmī dānishjūyān) was established at AMUT in 1970. Political pamphlets were distributed at different locations, such as the Students' Consumer Cooperative Store. Politically active students organized extracurricular activities such as mountain climbing. In the Iranian political culture, mountain climbing was deemed an important activity for a variety of physical, organizational, and psychological reasons. Climbing required a more trained body, assisted in nurturing a disciplined mind ready for collective activity, and was seen as an endeavor toward reaching a physical and metaphorical summit, where new horizions come into view. At another level of organizational networking, mountain climbing groups from various campuses were in contact with each other. This activity transformed the mountains into a training environment—a site of pedagogy and political organizing that was relatively much safer than the campuses, which were constantly under the surveillance of authorities. Maryam, a female alumni of AMUT who had participated in such mountain climbing, recounts:

In the mountains, everyone would sing epic songs. This was a result of youth excitement. We would wake up at 4 in the morning, somewhere in the middle of the mountain, packing our equipment, singing, this all created an atmosphere of seriousness, excitement. Many of the people who attended the university in the same year with me, for instance Taher Khorram, [who] was killed, was a peer of mine, he was studying mechanics, and used to attend mountain climbing. After a while, we noticed he was absent from the classes. (interview with Maryam, 2021)

Some dormitory rooms in AMUT were also involved in organizing student strikes (*Rūzigār sharīf* 2016, 96). Students from different political orientations had their own book and pamphlet collections and distribution networks. Islamic students, for instance, had a library at the prayer room where they collected books by Islamic oppositional thinkers and Muslim revolutionary theorists, such as Mehdi Bazargan (1907–1995), Ali Shariati (1933–1977), and Morteza Motahhari (1919–1979[6]; *Rūzigār sharīf* 2016, 97). Leftist student groups also had their own libraries, which included works by Amir-Hosein Ariyanpour (1925–2001), a leftist sociologist and translator, and Mikhail Sholokhov (1905–1984), a Soviet novelist. While these libraries could not officially hold the majority of banned Marxist texts, students would covertly pass them around despite regulations and restrictions. Maryam describes the situation like this:

We had a student library, an extracurricular library, this was different from the central Library of the university. The central library had many of the classics, such as Dante, etc. But our student library mostly had books by Ariyan Pour, Sholokhov, books of this sort . . . It was a library administered by students, books were also collected by students, per their taste. Whatever books that were not banned. The banned books were not placed on the shelves. The banned books would have been passed to each other but not put on the shelves. (interview with Maryam, 2021)

The demands of students encompassed both educational as well as broader political critiques, sometimes creating a tension between the two sets of demands. As we will discuss later in this chapter, the "educational" and "political" demands were at times formulated as a binary, even in opposition with one another.

Education-related concerns included issues of food quality, inadequate educational and pedagogical equipment, and dissatisfaction with grading scales and attendance requirements. One of our interviewees recounted how students organized a demonstration in support of a fellow student who faced the prospect of being dismissed from their studies for failing to meet certain grading requirements (interview with Ahmad, 2021). Omid Behrang, a student activist of the time at Polytechnic University, provided context in the ways in which strict grading measures impacted student activism at AMUT:

In early years, AMUT political atmosphere was lower than University of Tehran and Polytechnic University. Compared to University of Tehran and Polytechnic University, AMUT disciplinary rules were more difficult in course requirements, and grading. Students were expelled for the slightest violation. The courses were planned in such a way that the students had to attend the teacher's class all year round and study heavily in order to pass the final semester exams. However, this was not the case at the Polytechnic and I did not show up to the classroom many times and only studied on the night of the exam and easily passed the units. (Omid Behrang, personal correspondence, March 30, 2024 [translated from Persian])

Meanwhile, the political demands of students encompassed a broad range of issues, including socioeconomic concerns such as protesting against rising bus fares, and political issues like protesting against the killing of militant and political activists, calling for the immediate release of political and student prisoners, and criticizing visits by US officials.

Although there had previously been protests related to student-centered demands at AMUT, the first major political demonstration took place on February 21, 1970. This protest was in solidarity with the working class and was directed against the government's decision to increase bus ticket prices. Students from several major universities in Tehran, including AMUT, University of Tehran, Polytechnic University, and the Institute of Higher Education (dānishsarā ʿālī) participated in the demonstration. After leaving their classes, students gathered on the AMUT campus and protested.

The police responded harshly with various aggressive interventions, such as using tear gas against the unarmed crowd. The demonstrations continued for the next two days (*Rūzigār sharīf* 2016, 97–99). Ahmad, an alumni from AMUT, recounted his participation in this protest:

> I remember that event clearly. I participated in an act of flipping a bus in the street of Azadi in Tehran. At the time, this street's name was Eisenhower. We set it ablaze. City buses were painted red at the time. We shouted: "The red color of the buses is from the blood of the people" ("*Surkhī rang-i vāhid zi khawn mardumān ast*"). The guards attacked, and we fled. I went back the next day and noticed that the university was closed, and guards were stationed. The fact that they had shut down the university would have resulted in us gathering and demonstrating on the streets. (interview with Ahmad, 2021)

SIAHKAL INCIDENT: THE RELATION BETWEEN MILITANT ACTIVISM AND STUDENT ACTIVISM

On February 8, 1971, the infamous Siahkal incident occurred. Inspired by the guerrilla movement in Latin America, thirteen militant activists attacked a gendarmerie post in Siahkal, a village in the north of Iran. While the operation was a failure in the sense that all participating militants were arrested, convicted, and executed, Siahkal marked the beginning of armed struggle against the Pahlavi regime. As put by Vahabzadeh: "Siahkal became a 'resurgence' (*rastākhīz*) and an endless source of inspiration for a generation of dissenting Iranians" (Vahabzadeh 2011, 86). In their organizational accounts of the history of the student movement, both the OIPFG and the PMOI have discussed Siahkal as a decisive moment. According to *A Message to Students* (payām dānishjū), the student bulletin of OIPFG, the Siahkal incident was the turning point for the student movement, for it pushed students to think and act beyond the confines of the campus (*Payām-i dānishjū*, issue 1, 1975).[7]

Militant activism was thought about in theoretical terms by university students and alumni and, in turn, had immediate and profound effects on various campuses throughout Iran. The relationship between militant and student activism was two-way, organic, and complex. Guerrilla groups urged active participation by students, since students were considered the most knowledgeable and dynamic group in society. Indeed, some campuses became breeding grounds for militant organizations, where they distributed pamphlets, recruited new members, and mobilized and coordinated student forces in secret.

In fact, several members of the two most prominent militant political organizations of the time were students, faculty members, or alumni from AMUT. Among the members of PMOI, Majid Sharif Vaghefi (1948–1975), Morteza Samadiyeh-Labaf (1946–1976), Bahram Aram (1949–1976), and Vahid Afrakhteh (1951–1976) were students at AMUT, and Ali Bakeri (1943–1972) was a former assistant at the chemistry laboratory at AMUT (*Rūzigār sharīf* 2016 96). Majid, the brother and comrade of Masoud Ahmadzadeh, one of the theoretical leaders of militant activism in Iran, was also a student at AMUT and is considered by the OIPFG to have been the first AMUT militant student "martyred" by the security forces of the Shah.

Moreover, once OIPFG emerged in April 1971, it found many student members and sympathizers at AMUT.[8] According to OIPFG, Marxism-Leninism, as a science that deals with social formation and phenomena, is one of the most complex and layered subjects in nature and, therefore, requires a highly developed mind to understand. Due to their social background, academic studies, and psychological training, students were deemed capable of grasping and applying this theory.[9] OIPFG urged active participation by students to mobilize against imperialism and dictatorship, as they were considered the most knowledgeable and dynamic (āgāh-tarīn va mutiḥarik-tarīn) group of people within Iran (*Payām-i dānishjū* 1975, 14). Therefore, the organization aimed to mobilize and coordinate student forces. A document titled *A Message to Students* (*Payām-i dānishjū*), published in 1975 by this group, aimed to coordinate student activism in Iran, forge theoretical unity in student activism, and share experiences between student clandestine cells (*Payām-i dānishjū* 1975, 4). *Payām-i dānishjū* emphasized the decisive role of students in expanding the movement among the masses and asked them to fight back against imperialism and the dictatorship of the Shah through armed struggle (*Payām-i dānishjū* 1975, 40). OIPFG encouraged students to view their educational demands as a starting point to increase their political consciousness, and the student cells were tasked with politicizing student movements and creating connections between student movements and other populist movements. Students were also urged to promote their ideology, recruit new students, create ideological discipline in their political activity, and fight against the regime's cultural conspiracies. To accomplish this final task, students could use public and legal activities to increase their scientific knowledge. OIPFG also suggested boycotting the regime's cultural initiatives or exposing its cultural objectives (*Payām-i dānishjū* 1975, 55).

Radical students sought to extend their educational demands to include political ones. The militant activists also spoke in terms of the larger theoretical implications of their demands. For instance, PMOI argued that the most important objectives of the student movement were to become "more political," "more organized," and "more nationwide." In becoming more political, PMOI narrated how slogans "developed" from mere mundane student-based demands such as "free education" into more sophisticated demands such as "autonomy and political freedom at universities," the "right of students to have a share and supervise the appointment of chancellors and deans," and political slogans against imperialism and the "White Revolution." Furthermore, it was understood that if they were to become a nationwide movement, intercampus relations and broad solidarity would be necessary (PMOI, *infijār maqar gārd dānishgāh ṣan'atī* [Blowing up the Campus Police Station at San'ati University], 1974).

The portryal of of student demands for free education or autonomy of the university as belonging to an early, "lower" political stage points to dogmas that characterized certain elements of the radical left. We believe that a more nuanced formulation would reveal that students' educational demands were both sophisticated and deeply intertwined with their broader social and political views. At AMUT, the pedagogical aspirations and demands of students and faculty were interwoven and interconnected with their anti-Shah and anti-imperialistic politics.

As the militant movement gained momentum, two seminal texts authored by Marxist-Leninist-Maoist theorists of armed struggle, Amir Parviz Pouyan and Masoud Ahmadzadeh, came to be highly regarded among students in Tehran, including those at AMUT. Both authors were inspired by the role of vanguard intellectuals in the Latin American movement and advocated for the same tactics for Iran (Rahnema 2021). An examination of these two pamphlets will aid in comprehending how early theorists of the militant movement depicted Iranian society and why they advocated armed struggle as the sole feasible political solution.[10]

THE POLITICAL IDEOLOGIES OF OIPFG

In spring of 1970, Amir Parviz Pouyan authored his pamphlet, *The Necessity of Armed Struggle and Refuting the Theory of Survival*. In this work, he portrayed the coup of 1953 as the defeat of anti-imperialist movements in Iran, leading to the rise of a fascist regime representing imperialism. Pouyan described the Iranian social system as being policed by a fascist state and argued that revolutionary intellectuals had no established connection with the masses (Pouyan 1970, 24). He further argued that while the most violent labor exploitation was happening in both governmental and private factories, the political suppression and constant monitoring of workers by SAVAK had resulted in political apathy among workers and a submission to a *petit bourgeois* culture that the regime had successfully implanted in the minds of the workers. Pouyan argued that the main reason the exploitative relations had not resulted in revolutionary consciousness was that workers viewed the strength of their enemy as being absolute and, thus, perceived their own ability to emancipate themselves from this dominance as absolutely nonexistent. This statement, commonly described as "the two absolutes" argument, became the backbone on which he put forward his militant solution.

According to Pouyan, the *only* way to get beyond this absolute weakness—to politicize and organize the working class—was for the revolutionary intellectuals to connect with the masses through "revolutionary might" (*qudrat 'inqilābī*). He contended that such revolutionary might would establish emotional bonds between revolutionary intellectuals and the proletariat, eventually extending into building organizational relations (Pouyan 1970, 26). Consequently, Pouyan suggested the necessity of armed struggle headed by the revolutionary intellectuals, for such a struggle would have a propagandistic nature, informing the proletariat of a force of its own, shattering the idea that the power of their enemy was absolute. Revolutionary violence would serve two functions: First, it would help build the necessary progressive consciousness in the proletariat, and second, it would encourage the proletariat to actively play a role in building the common future.

Pouyan was critical of the established Marxist-Leninist organizations of Iran, most specifically the Tudeh Party of Iran, for insisting that rebels must wait for "a right moment" or "desirable condition" to activate revolutionary struggle (Pouyan 1970, 32). In his opinion, this exhortation to wait for the perfect opportunity was a submissive and opportunist approach. He further argued that such "a right moment" was only a metaphysical concept, which was used as a tool to conceal the apparent weakness of the theory and strategy of these organizations. He wrote, "The right moment and desirable condition will never be actualized unless revolutionary elements respond correctly to the historical necessities" (Pouyan 1970, 33). In his final remarks, he noted that the revolutionary cells should be covert, while also actively and aggressively using revolutionary might to attack the enemy.

The second key theorist of armed struggle with wide readership among students was Masoud Ahmadzadeh. In the summer of 1970, he wrote his famous pamphlet, *Armed Struggle: Both a Strategy and a Tactic*. Compared to Pouyan's pamphlet, Ahmadzadeh's treatise is a longer argument, going into much more detail in terms of its sociological description of Iran. Yet even so, Ahmadzadeh shares similar arguments to Pouyan about how revolutionary change should proceed. The foundational historical and sociological analyses of the two texts are similar. In the same manner as Pouyan, Ahmadzadeh also described the coup of 1953 as an imperialist one, which resulted in the disintegration of all national and anti-imperialist political organizations, and was similarly critical of the theory and practice of the Tudeh Party of Iran (Ahmazadeh 1970, 17).

Ahmadzadeh claimed that the objective of the White Revolution was for comprador capitalism[11] and bureaucratic capitalism to extend its economic, political, and cultural dominance in the villages (Ahmadzadeh 1970, 22). To support this statement, he argued that as a result of the White Revolution, the contradiction between comprador feudalism and imperialist interests was resolved through suppressing feudalism. Therefore, the contradiction that peasants in Iran were experiencing was not with feudalism but with imperialism and its dependent governmental machinery (*māshīn dawlatī vābastah*; Ahmadzadeh 1970, 27). Consequently, Ahmadzadeh concluded, following the White Revolution, that all the internal contradictions in Iranian society were overshadowed by one primary contradiction: the contradiction between oppressed people and Western imperialism. Consistent with this mode of theorization, OIPFG also encouraged students to put aside all other social contradictions and focus all of their efforts on eliminating this main contradiction (Ahmadza-deh 1970, 9).

As a result, there was rarely any discussion of other social issues, includ-ing women's rights, in most of the texts produced by student and militant organizations at the time. Theoretically, this implies that both PMOI and OIPFG believed that all these other social contradictions would automati-cally be solved once the main social contradiction had been resolved. We argue that this is an example of the violence of abstraction (Sayer 1987 when a theory and strategy of social change violently eliminates compet-ing social demands in favor of one, rather than delicately conceptualizing the relations between all the various demands.

Compared to Pouyan, Ahmadzadeh elaborated more on the interna-tional split in the left movements and went into some discussion of the debates between Marxism-Leninism and revisionism in the Soviet Union. He argued that in this split, the true haven for revolutionaries was Marxist-Leninist-Maoism. In this condition, new communist groups were estab-lished in Iran, which were informed by Mao's books, yet departed from the Chinese revolution because they were drawing influence from the Cuban revolutionary practices as well. Ahmadzadeh suggested that a key ques-tion faced the Iranian communists: Should they move toward establishing a proletariat party in the traditional Marxist-Leninist or Maoist sense, or should they establish armed cells in villages and initiate guerrilla warfare, as exemplified by the case of Cuba? (Ahmadzadeh 1970, 19).

Ahmadzadeh noted that a socialist consciousness should be passed on to workers through building relations between intellectual circles and workers. Even so, no such active connection existed in Iran. Moreover, the imperialist dictatorship had coercively suppressed every kind of rev-olutionary movement, while aggressively sponsoring the dissemination of political and ideological propaganda (Ahmadzadeh 1970, 37). Under such conditions, Ahmadzadeh, just as Pouyan had done, emphasized the importance of vanguard and professional revolutionaries, who would lead the guerrilla warfare, which he deemed essential to organize the masses (Ahmadzadeh 1970, 42). Informed by Regis Debray's *Revolution in the Revo-lution?* (1967), Ahmadzadeh concluded that it was only the "small armed engine" that could activate "the larger engine" of the masses (Ahmadzadeh 1970, 67). Ahmadzadeh's pamphlet ends with a call for the revolutionary groups to instigate their political-military campaign in any location that they deem crucial.

As clear from the previous discussion of these two texts, notwithstanding certain minor differences, both Pouyan and Ahmadzadeh shared much in terms of how they depicted and historicized Iranian society and the way they formulated the necessary anti-imperialist praxis ahead.[12] Both these texts were commonly read by student activists, and their main arguments were the constant subject of critical evaluation by the Tudeh Party's student publications, such as *Paykār*.

Paykār, the student bulletin of the Tudeh Party, was first published in 1971. It was meant as a vehicle for the Tudeh Party to remain active among students. The objective of this journal was to distribute Marxism-Leninism among students and combat any kind of "deviation" in student activism. Contending that the political struggle should be centered around communist parties and avoid militant strategies at the time, by "deviation," this journal referred specifically to the role that militant leftist groups were assigning to vanguard intellectuals, the influences of Maoism among students in Iran, as well as in the Confederation of Iranian Students abroad. Calling them semi-leftists (*chap raw, chap namā*) who have deviated from the correct path and are pressing on a path which does not correspond with the current historical conditions (*sharāyiṭ mushakhas kunūnī; Paykār* 1971, issue 1, p. 21).

Paykār also emphasized the necessity of legal and public activity (*fa'alīyat 'alanī wa qanūnī*) for student organizations. It argued that the student movement should not be mistaken for political parties. Student organizations are "mass organizations" (*sāzimān tudah'ī*), not political ones. They should organize masses of students, notwithstanding their class relations, and not only the "scouts" (*pishāhang; Paykār* 1971 issue 1, 33). A student organization was not to follow a specific ideology but needed to organize students who might have different, even contradictory ideologies.

Nevertheless, the Tudeh Party argued that the student movement had a democratic anti-imperialist characteristic. According to the Tudeh Party, students in Iran were predominantly from the middle class, and these classes have anti-imperialist and antiregressive tendencies.[13] Thus, the objectives of the student movement and student organizations had to be to act as part of Iran's anti-imperialist and prodemocracy movement, but they also simultaneously needed to safeguard the interests of the majority of the students (*Paykār* 1971, issue 1, 33).

It is based on this latter characteristic that the Tudeh Party emphasized the necessity of public and legal activity for student organizations. To act legally, the student organizations had to work under the laws of the country and could not reject the constitution (*Paykār* 1971 issue 1, 33). In other words, the objective of a student organization was not to be social revolutionaries and try to overthrow the regime (*Paykār* 1971 issue 2, 19). Instead, the Tudeh Party strongly suggested that the student movement should call for the enforcement of the democratic rights and freedoms that the constitution of Iran had secured (*Paykār* 1971 issue 2, 19).

To fully grasp the significance of these texts and debates, which were extensively read on Iranian campuses, we must consider Iran's sociological conditions, the relationship between the Iranian state and the global imperial order, and the discussions surrounding pathways for social change. Student activists not only dramatized and popularized these ideas but also expanded upon and further theorized the concepts of militant anti-imperialist struggle and the role of armed propaganda. Ali Rahnema, a historian and scholar of militant movements in Iran, notes that:

The Iranian guerrilla movement, through its praxis, established a frame of reference, an ethos, and an archetype for Iranian political activists. It would be fair to say that its struggle and comportment established a code of conduct for the politicized youth. The battle conducted by the Iranian guerrilla movement captured the imagination of urban Iranians, especially its youth, and confronted them with important political questions on how to engage with authoritarian rule. (Rahnema 2021, 4)

Clearly, we can trace the effects of the rise of armed struggle on the campus and among student activists as well. On the AMUT campus, after the Siahkal incident, engineering student activists grew increasingly political in their demands, expanding their concerns to encompass the working class.

The leaders of the militant activism movement embraced Marxism-Leninism but did not align themselves with either Moscow or Beijing or as the representatives of Marxist orthodoxy. These leaders, including Amir Parviz Pouyan and Masoud Ahmadzadeh, considered themselves independent Marxists and criticized the Tudeh Party, the oldest Communist Party in

Iran, for its "gradualist and mass mobilization tactics." Instead, they advocated for armed struggle. According to Dorraj, most of the recruits to these organizations came from educated middle-class youth, particularly students, educators, and engineers (Dorraj 2006).

Janet Afary has observed that OIPFG members primarily came from "the modern middle class," whereas the PMOI tended to recruit more from "the old middle-class sectors, where Shi'i rituals remained an integral part of daily life" (Afary 2009, 245; also see: Abrahamian 1989). Despite some ideological differences, such as the PMOI's adoption of Marxist theory alongside Islamic traditions (as opposed to OIPFG's secular Marxism), the two groups generally had a positive relationship (Vahabzadeh 2011). Some students at the time were also unaware of the distinctions between the two organizations before the revolution. One of our interviewees stated:

I only understood the difference between Mujahidin and Fadayian, once I graduated from the university. That these are two separate organizations with such and such [organization and ideological] differences. I didn't know about that during my student days. (interview with Ahmad, 2021)[14]

The close relation between the leftists and Islamic leftists have also been discussed by one of our female interviewees: "The two groups were not enemies to each other, would cooperate with each other and had many similarities with each other. The leftist girls also wore simple clothes, so were respected by the religious" (interview with Batol, 2021).

The relative lack of ideological conflict between the two groups has been criticized by Vahahzadeh, arguing that "the Unity of action" resulted in the collapse of the distinction between OIPFG and PMOI militants. Vahabzadeh states: "In the context of an increasingly polarized society, this so-called unity of action through armed struggle gradually clouded the differences between the Left and the religious groups (a difference that must be maintained at all times) and led to the Left's forgetting its own secular foundations" (Vahabzadeh 2011, 92).

With the developments within OIPFG and PMOI, and the proliferation of leftist and left Islamist groups in the final months before and early days of the revolution, AMUT students ended up in various political organizations. Shiva Farahmand Rad, the exiled author we mentioned earlier, not only depicted the political and cultural history of AMUT students in his reflective pieces but also created a list of 135 AMUT students and alumni who were

killed during the revolutionary years, up to 1985. This list includes information on the years of attendance at the university and their organizational
affiliations. Notably, the list reveals that while OIPFG and PMOI, the two
prominent militant organizations at the time, had members from AMUT,
the political diversity of AMUT gradually extended well beyond these two
groups. The great differences in political affiliations and levels and modes
of sympathy with OIPFG and PMOI among our interviewees also affirms
this same general interpretation. It is beyond our scope to delve into the
complex landscape of leftist groups at the time, including the nuances of
their theoretical and organizational divisions and developments. However,
it is important to acknowledge the continued activism of several AMUT
students in these groups as they expanded over the years, especially in the
immediate years before and after the revolution.[15] It is not the objective of
this study to discuss the theoretical debates and practical implications that
have been the basis of this proliferation of leftist and communist political
groups. It suffices to mention that despite their shared revolutionary stance
against the Shah, these organizations held divergent ideological and strategic positions. The involvement of students in these groups underscores
that, particularly in the late 1970s, nuanced political and organizational
distinctions gradually emerged among leftist students at AMUT, leading to
diversions and tensions over time.

KEY STUDENT DEMONSTRATIONS AT AMUT
AFTER SIAHKAL INCIDENT

Following the Siahkal incident, on May 2, 1971, students from AMUT
staged a demonstration to show solidarity with the ongoing protests at
the Engineering Faculty of the University of Tehran. A report featured in
the *Central Intelligence Bulletin*, produced by the Directorate of Intelligence
in the US and published on May 5, 1971, provided a detailed account of
these demonstrations and noted the growing influence of Marxist ideology
among Iranian student protests.

The report stated that a week earlier, students at the University of Tehran
had protested the arrest of seven of their fellow classmates in connection
with demonstrations that had taken place in November and December
of the previous year. After a meeting at the University of Tehran, which
was intended to foster dialogue between students and faculty, was broken

up, approximately 700 students began a march on campus. They chanted slogans identified with Marxist causes and criticized the Shah, his reform programs, and the expenses associated with Iran's twenty-fifth centenary celebrations. The police responded with force, in contrast to their restrained response during similar protests during the previous year, and used riot clubs, tear gas, and submachine guns—an automatic firearm that is fed ammunition through a magazine and designed to shoot pistol cartridges—to disperse the crowd. Approximately 250 students were arrested, and an equal number were injured, some of them quite severely. As a result, police were stationed on campus for the first time since 1968 (*Central Intelligence Bulletin*, May 5, 1971, CIA-RDP79T00975A018900100001–1).

Over the weekend, demonstrations of solidarity spread to other colleges in Tehran, including AMUT. Riot police were stationed outside the Polytechnic and National Universities, and they entered AMUT to disperse the protesters. The students at AMUT were showing support for the students at the University of Tehran, protesting against the killing of striking workers at the Chit Jahan factory, and protesting the upcoming 2,500-year celebration of the Persian Empire. These celebrations were perceived as a series of state-sponsored propaganda events to commemorate the institution of the US-backed monarchy. Faculty members at AMUT joined the students, and approximately 400 students and several faculty members were detained. The following day, approximately sixty faculty members held a meeting, during which a physics professor, Farhad Ardalan (b. 1939), proposed that all faculty members collectively resign. The letter containing the signatures of the faculty members was passed along to the Iranian prime minister, Amir Abbas Hoveyda (who served between 1965–1977), who appeased the protesting faculty and promised to release the students. As a result, the faculty members returned to work (*Rūzigār sharīf* 2016, 108). This event marked an early instance of the dynamic and close relationship between student and faculty activism, which soon became a defining characteristic of the political environment at AMUT.

Nevertheless, the CIA was under the impression that such communist elements were "inspired or directed" from outside the country and would not become widespread among students:

The use of slogans identified with anti-government Marxist causes, which are known to be funded from Communist sources, reinforces earlier indications that among the protestors are hard-core leftists who the U.S. Embassy believes are inspired or directed from outside the country. As such, they are likely to remain inimical to the Shah and his programs regardless of any conciliatory gestures made by the government. The demonstrations appeared to have little support outside the core group of student agitators, however, and are believed to have had little effect on the student body as a whole. (*Central Intelligence Bulletin*, May 5, 1971, CIA-RDP79T00975A018900100001–1)

As subsequent events demonstrate, the CIA severely underestimated the force and intensity of Marxist and leftist groups on AMUT's campus.[16]

In his interview with the Harvard Oral History Project, Zarghamee, who later became the chancellor of AMUT, observed that after this incident, there was a growing sense of SAVAK surveillance, even among professors at the university, to the extent that certain faculty were feeling their tenureship and career advancement would be threatened by their activities. This feeling of censorship, surveillance, and disempowerment among faculty resulted in the establishment of the first Iranian faculty union, named "*Kumiteh rifāh*" [Welfare Committee] (Zarghamee, interview with HOHP; also see Mojab 1991). Critically, key professors within the union, such as Fereydon Hojabri, played leading roles in shaping the most eventful struggle for university autonomy in Iran's history (Hojabri, 2016).

The political demonstrations at AMUT reached the Ramsar Educational Conference discussed in earlier chapters. This conference, held annually in July and August, was led by the Shah and focused on discussing government policies and achievements in implementing the twelfth principle of the White Revolution, also known as the "Educational Revolution." The fourth conference specifically emphasized the establishment of "law and order" (*ārāmish*) on campuses. The conference declaration called for the purification (*sālim sāzī*) of educational organizations and criticized the lack of public awareness about "the essential concepts of the revolution." The fourth article in the declaration noted that universities and institutions of higher education in Iran would no longer tolerate activities deemed antinational and anti–White Revolution (*zid millī, wa zid inqilāb īrān*; *Paykār* 1971, issue 2, 6).[17]

These discussions led to the establishment of the University Guards, which were essentially an armed university police force responsible for monitoring and intervening in student demonstrations. The establishment of the University Guards in the early 1970s marked the first step toward the militarization of the campus.[18] However, the establishment of the guard also resulted in the further politicization of the students. The militant activists now saw them as the personification of the state on the AMUT campus. Resistance to the campus guards was seen as resistance to the regime, and an attack on the University Guards was seen as an attack on the state as a collaborator with US imperialism. Throughout the 1970s, several clashes between the University Guards and AMUT students were observed. The police guard at AMUT was subjected to two acts of terror at the time. The first one was by PMOI, who blew up the Guard station, followed by a second attack a few years later by the OIPFG, who killed the head of the AMUT Guard, Sarvan Yadollah Nowruzi.[19]

In early 1970s, we can see student demonstrations following specific human rights violations and executions,[20] against visits by US officials,[21] in protest against the killing of students,[22] and in protest against the admission of military students,[23] as well as on several other occasions. Every year, in early December the annual celebration of Student Day loomed large. This time of year was the high season for activism, and the most common student slogan, "Unity, Struggle, Victory" ('itihād, mubārizah, pirūzī), echoed through the campus (Rūzigār sharīf 2016, 135).

As students began to employ new methods of activism, the university responded with new measures to curtail their efforts. By 1973, boycotting classes had become a common form of protest. One early example occurred on June 4, 1973, when University Guards attacked and detained fifteen students following a protest at AMUT. In response, students refrained from attending classes for several days. To counteract this strategy, the university council issued an ultimatum on June 9, stating that students must attend classes, or their summer semester would be canceled (ḥazf kardah; Rūzigār sharīf 2016, 136).

NEW MEASURES, NEW RESISTANCE: THE EVOLUTION OF UNIVERSITY POLICIES AND STUDENT ACTIVISM

During the 1973–1974 academic year, the new university chancellor, Hossein Nasr, implemented several initiatives, including the introduction of a computerized system to track student attendance and automatically disenroll those who exceeded a certain number of absences (Rūzigār sharīf

2016, 143). In addition, SAVAK began categorizing politically active students as provocateur groups 1 and 2 (*muḥarik*). Being labeled provocateur 1 resulted in the student being deprived of student privileges (*'imtiyāzāt dānishjūʾī*), while being labeled provocateur 2 led to the loss of dormitory privileges (*Rūzigār sharīf* 2016, 146).

The student protests for Student Day in 1973 began on November 27. Within ten days, 200 students were prevented from finishing their first semester due to their participation in the protests. Despite this, the students persisted with their demonstrations. On December 17, one-third of the entire student body, approximately 1,200 students, were disenrolled from their first semester (*Rūzigār sharīf* 2016, 148).

Starting in January 1975, students initiated a new form of protest by boycotting final exams. During the exam sessions, students would attend but intentionally leave their answer sheets blank. In response, the university administration adopted various tactics to pressure the students. For instance, University Guards marked the student cards of those they classified as provocateurs. Nevertheless, in a show of solidarity, students punched their own cards, blurring the distinction between cards punched by the Guards and those punched by the students themselves. As a result, many exams were canceled. This pattern continued, and on the eighth day, none of the 500 students showed up for the exams. The same pattern repeated on February 1 and 2, 1975 (*Rūzigār sharīf* 2016, 153).

The university attempted to quell this new form of activism by imposing a rule that allowed a student to enroll in only as many courses the next semester as they had passed in the previous semester. As a result, the second semester was relatively quiet, with less noticeable student activism. It was during this comparatively calm semester that an alleged assassination planned by PMOI on Hossein Nasr took place, which led to his resignation (Nasr 2014; *Rūzigār sharīf* 2016, 156). According to Nasr, his tough stance against student activists was the reason he was included in PMOI's list of terrorist targets. Nasr notes:

The decision to use force is sometimes unavoidable, especially when the other side uses force, as the Quran itself teaches. I mean, if you have students breaking windows and throwing television sets out of their rooms, which they were doing . . . What were you going to do? (Nasr 2014, 123 [Translated from Persian])

VIOLENCE AGAINST CAMPUS POLICE

As noted, in 1975, AMUT was the scene of a few violent actions by militant opposition groups, especially the PMOI. PMOI was active at AMUT during this period and played a crucial role in specific events throughout the history of the university. One of PMOI's active members, Ali Bakeri, who worked in the chemistry laboratory and had connections with regional militant organizations, was executed by the Pahlavi regime on April 19, 1972. The execution aroused agitation among students and chemistry professors, and some students even called for the chemistry department to be renamed in his honor (*Rūzigār sharīf*, 2016). The anniversary of Bakeri's execution continues to be a time of activism and unrest on campus.

On April 20, 1974, the second anniversary of the execution of Ali Bakeri, PMOI planted a bomb on the roof of the campus police station at AMUT. Although the explosion did not result in any casualties, it was a significant example of continuing efforts to terrorize and demoralize the police forces during that era. The bombing of the guard station by PMOI was the most noticeable event during this otherwise uneventful semester at AMUT. A pamphlet titled *Blowing up the Campus Police Station at San'ati University*, written by PMOI, was published in 1974, just a few days after the incident on April 20. Soon after the bombing of the Guard building, Nasr, afraid that he was on the list of potential targets of terrorist attacks of this organization, resigned from his position.

The waves of student activism between 1971 and 1977 were notably influenced by developments within the militant movement in general. Thus, a short note on these developments will be helpful. Starting in the winter of 1974, an important ideological shift within the PMOI became noticeable, which resulted in publication of the *Manifesto on Ideological Issues*, in which the central leadership declared "that after ten years of secret existence, four years of armed struggle, and two years of intense ideological rethinking," the organization was shifting from its earlier belief in Islam and its social base in Iran's "*petite bourgeoisie*" to a more forceful Marxist philosophy and reliance on Iran's proletariat. The internal ideological tensions escalated to the point where they led to the expulsion of half of the members of the organization,[24] as well as the killing of certain people among its leadership by the organization itself, including Sharif Vaghefi, a former AMUT student. In the postrevolutionary period, the name of AMUT was changed to Sharif University, in an attempt by the Islamic state to replace Aryamehr (linked to the Shah) with the name of a former student activist known to have had Islamic sympathies.

While this move arguably dissolved the ideological distinction between the PMOI and the OIPFG, because of the OIPFG criticism of the PMOI leadership in their handling of the ideological shift and the quarrels and differences over tactics and leadership principles between the organizations, the two groups never merged or formed a unified front.[25] Within a year, the leadership of both organizations was significantly weakened because SAVAK managed to kill important figures in both groups.[26] Consequently, in the case of the OIPFG, a tactical shift was introduced. Vahabzadeh notes, "The cherished notion of guerrilla warfare was questioned, and the 'militant' theory of Ahmadzadeh began losing authority and was gradually replaced with the 'political' theory of Jazani." This shift, in practical terms, meant that the OIPFG moved to establish a political wing and shifted toward "organizing workers and students in nonmilitant cells that would lead spontaneous protest movements" (Vahabzadeh 2011, 87). Following these events, both groups remained relatively weaker, and this remained so until early 1978, when, in the context of the rise of the revolutionary movement, membership in both groups was replenished and significantly increased. As noted by Vahabzadeh, "the heyday of the guerrillas began to fade in 1975. The indifference of the people and decline of student support for the armed movement shattered the zealous optimism of the Fedayee militants about the perceived imminent revolution" (2011, 87).

Against their expectations, the militant activist groups were failing to spark broad political protests, and from an organizational standpoint, their covert leadership had been severely undermined by the SAVAK. We can see that as the militant movement faced such organizational and setbacks, by 1975, both POMI and OIPFG had lost their ability to act as an organizing force on the campus. A report from the US embassy in Tehran to the secretary of state in Washington describes the student demonstration in December 1976 by saying that the American mission has been "pieced together" from government of Iran officials, faculty members, students, and administrators.

This report observes that "demonstrations this year were more wide-spread, less directly political than in the past, and more intelligently managed by university administrators." The sequence of protests began that year in the first week of October when revised government fee and loan policies cut back support for some students, leading to student strikes in nearly all universities. As a result, AMUT had only been intermittently open since that time. This letter ends with the following statement:

While publicly stated goals of dissenters related to fees, food, and dormitory inadequacies, some faculty members who have significant student contact feel absence of apparent political motivation was result of tactical choice of radicals who have political goals but prefer to avoid giving GOI excuse to bring in heavy security reinforcements. At least some students agree. Behavior of police this year not as extensively criticized by faculty as in past, but as one professor said, "perhaps we are just getting more accustomed to our annual December invasions." (Letter from American Embassy in Tehran to Secretary of State, Subject: Annual Student Brouhaha: Widespread but less violent, December 1976)

By November 1977, the student movement had gained significant momentum and appeared to be on the brink of imminent action. On November 16 of that year, a statement written by a SAVAK agent said, "In these days any second there might be a student demonstration (*dar in rūzha har laḥzah imkān tashanuj dānishjū'ī vujūd dārad*)." The agent reported on many students at AMUT who demonstrated on Eisenhower Street, saying they started to "break windows of banks and stores" and were aiming to join students at the University of Tehran.[27]

AMUT SIT-IN: A PIVOTAL EVENT IN STUDENT ACTIVISM HISTORY

In late October 1977, a few days after the Goethe Poetry nights,[28] the Student Group for Cultural Research at AMUT (*gurūh-i dānishjū'ī pazūhish-hā-yi farhangī dānishgāh ṣan'atī*) organized a weekly lecture by a member of Iran's Writer Association at AMUT in collaboration with Iran's Writer Association. The third lecture was on November 19.[29] On that night, according to the *Guardian*, "eye witnesses [sic] said that riot police attacked a crowd of 2,000 gathered outside the university." A total of fifty people were initially arrested, and fourteen were later released. A minimum of thirty-nine students were injured. An audience of 4,000 were already inside the campus. The lecturer, prominent leftist poet Saeed Soltanpour, who was going to present on the subject of "theater and freedom," protested the arrests by refraining from presenting. Student organizers of these events asked attendees to stay at the hall, resulting in a night of a sit-in strike. Approximately 4,000 attendees stayed at the hall over the night (*Guardian*, November 17, 1977).[30]

Significantly, these dramatic events coincided with the Shah's visit to the US during the Carter presidency. Because of Carter's outspoken official rhetoric of protecting human rights, many assumed the Shah and his regime would not engage in any flagrant human rights violations during his visit. However, according to Liz Thurgood, reporter for the *Guardian* in Tehran, the brutal crackdown on dissidents continued unabated for ten days, after which "hundreds of intellectuals, former politicians, and students" were "nursing wounds received under the truncheons of riot police, and in a new twist, what apparently were trained attackers" (*Guardian*, November 17, 1977).

The following afternoon, having been promised that the detainees would be released, the attendees came out of the hall and arrived at the street, already filled with crowds who had come to display their support and solidarity with the protesters. Demonstrations continued, and 150 people were injured. The protestors chanted: "Workers are our brothers/ Brothers should be Equal" (*kārgarān barādaran, barādarān barābaran*); "Shah is United States' Chained Dog" (*shāh sag-i zanjīrī āmrīkāst*); "Brotherhood, Equality, Workers' Government" (*barādarī, barābarī, ḥukūmat kārgarī; Rūzigār sharīf* 2016, 173).

Omid Behrang, a former student activist who was directing the student-led library at Polytechnic University at the time, shared some of his reflections on the significance of the November 1977 sit-in at the AMUT event in a personal correspondence with us. His insights help illustrate the interconnectedness of student activism across several universities at the time, as well as the role of this specific event at AMUT for student activists in general. We quote his correspondence here at length:

The overnight sit-in at the University of Technology was actually a sign of an important shift in the political spirit of the student movement. It was one of the manifestations of the onset of the 1979 revolution. Additionally, it was a sign that the student movement had expanded significantly. This movement was no longer limited to just a few dozen activists at each university. To my knowledge, several hundred people (perhaps more) participated in that sit-in, and the following day, students from all universities moved individually and collectively to join the sit-in. Personally, I was near the University of Technology when I saw the guards attacking people on a large scale around the university. Years after the mass demonstrations related to Gholamreza Takhti death and the mass protests against the high price of bus tickets, Tehran once again witnessed a massive student movement. Geographically, Eisenhower Street—later renamed Azadi—was the scene of the Guard's attack and the deployment of tear gas. Until this period, student activists were constantly trying to take their demonstrations to the streets, but that day it was the Guard that facilitated the geographical expansion of the student movement. (Omid Behrang, personal correspondence, March 30, 2024 [translated from Persian])

A report by the *Washington Post* provides some insights into the way police suppressed the movement on that day. Police altered their tactics by utilizing mobs to confront student dissidents, possibly to portray incidents as clashes between "unruly extremists and patriotic pro-Shah citizens" Branigin, William. "Shah's Opponents Beaten by Mob in Iran." *The Washington Post*, November 22, 1977. William Branigin's report for the *Washington Post* revealed that the mobs consisting of 100 men, dressed in civilian attire, arrived in two green buses and physically prevented him and other Western journalists from following the dissidents, who were being chased away from AMUT. Despite the arrival of two truckloads of riot police, the mobs, who were each carrying identical "rough-hewn clubs about two feet long," were left untouched (*Washington Post*, November 22, 1977).

THE QUEST FOR AUTONOMY AT AMUT IN THE LEAD-UP TO THE 1979 REVOLUTION

In the months just prior to the 1979 revolution, activism at AMUT focused on one specific key demand: autonomy from the state. Scholar of Iranian history and education Shahrzad Mojab explains that AMUT had some elective bodies, such as the University Council, that created conditions that made opposition to further court and state restrictions on the university plausible (Mojab 1991). One of our interviewees, Professor Ahmadiyan [name psyudonymized], also emphasized that councils (*shurā-hā*) were an important part of the university that contributed to maintaining a degree of autonomy. Mojab explains, "While Aryamehr University, like other institutions, was not autonomous (the president, vice-presidents and supervisors of administrative office were all appointees), it did enjoy more self-governing status on the college level than other institutions" (Mojab 1991). For example, members of the councils of the faculties and departments were given the authority to elect the deans of faculties, as well as the heads of research and instructional centers. This was different from other Iranian universities at the time, where the deans were appointed rather than elected by an elected council.

In early 1977, the faculty of AMUT protested the government presence on the campus. Since the University Council did not suppress their protest, the government under Prime Minister Hovayda declared the transfer of the university to Isfahan. This was against articles 1 and 2 of the charter of the university (interview with Mohammad Ali Ranjbar, published in *Sharīf az āqāz tā kunūn bi rivāyat ru'asāy-i ān* 2006, 40). Because the transfer would result in the closure of the Tehran campus, it was also contrary to goals of the country's sixth developmental plan, which had advised that Iran would need 60,000 new engineers within five years (at the time, AMUT graduated 700 new engineers each year).[31]

On July 6, 1977, Zarghamee, chancellor of AMUT, announced that starting in the next academic year, the Tehran campus of AMUT would no longer accept new students for three years, and the Tehran campus would be administered under the Isfahan campus (*Rūzigār sharīf* 2016, 168). Upon transferring the university to Isfahan,[32] the government would convert the facilities of AMUT into a Center for Military Sciences and Arts, which would undertake the responsibility of training military experts. At the same time, twelve faculty members were sent to work in clerical and governmental positions (interview with Ranjbar, published in *Sharīf az āqāz tā kunūn bi rivāyat ru'asāy-i ān* 2006, 41).

From July 1977 to February 1978, negotiations took place between faculty, governmental authorities, and members of the board of trustees. In September 1977, the previous chancellor was dismissed and replaced by Alireza Mehran,[33] and the faculty members who had been sent away were asked to return. In December, around the annual Student Day demonstrations, students started protesting against the regime and the dissolving of the Tehran campus of AMUT. Chancellor Mehran threatened that if the students did not return to classes within the next three days, the Tehran campus would be shut down for a year (*Rūzigār Sharif* 2016, 173).

On January 14, 1978, it was announced that the university was to be shut down immediately until further notice. In February 1978, the negotiations failed. The aim of the faculty was to stop the dissolution of the university. A letter signed by 138 faculty members was sent to Prime Minister Amuzegar, asking him to reverse the decision and not to relocate the campus. However, this letter received no response from the prime minister (*Rūzigār sharīf* 2016, 176).

On April 30, 1978, the university management blocked students from entering the campus. The professors pointed out that the charter of the university clearly stated that the permanent location of the university would be in Tehran. They argued that according to the charter, the Isfahan campus could be a branch of AMUT, not its primary campus Kumīteh Intishārāt Dānishgāh. Būltan Khabarī. issue 2, 27 Ordibehesht 2537. On May 7, 1978, after more than a year of failed negotiations, the teaching staff, inclusive of full-time faculty and instructors, ratified a resolution asserting that they would not continue performing any educational activities until the official acceptance of new students (*Rūzigār sharīf* 2016, 176).[34] On the same day, the university management shut down the lunchroom of the university staff. Vahid Khansari (a student at AMUT at the time) argues that the AMUT strike against the government policy to move the campus to Isfahan was the first clerical strike (*I'tiṣāb kārmandī*)[35] in Iran. In a few months, other educational institutions and governmental organizations followed this lead. While demonstrations had happened in Tabriz and Yazd, no clerical strike occurred before the one by AMUT (*Sharīf az āqāz tā kunūn bi rivāyat 'asātīd* 2009, 60).

AMUT faculty were supported by other university activists and wider intellectual networks. The faculty at the University of Tehran wrote a letter of support,[36] stressing the national need for engineering expertise and the crucial position of AMUT in fulfilling this need. The faculty of the University of Tehran also emphasized that establishing a university in another part of the country should not be at the expense of shutting down another campus (*Būltan khabarī*, issue 2, 27 Ordibehesht 2537). On May 13, 1978, students at the National University (now called Shahid Beheshti University) expressed their solidarity with the striking AMUT faculty. Furthermore, students at Polytechnic University canceled classes as an act of solidarity (*Būltan khabarī*, issue 2, 27 Ordibehesht 2537).[37]

The striking faculty and students established the *News Bulletin* to publicize their demands as well as other pertinent information related to their strike. The requests of the faculty as expressed in this *News Bulletin* were: 1) Autonomy and freedom at the university, 2) respect for the university's regulations and laws, 3) opposition to inhumane violence at the university and the presence of the Guards on campus, and 4) the right to solidarity and to establish organizations (*tashkīl*) for following up on their legal demands (*Būltan khabarī*, issue 2, 27 Ordibehesht 2537). The government reacted by cutting the monthly stipend for each of the striking professors. The faculty did not surrender. Instead, they established a committee called the Committee for Emergency Loan (*kumītah vām iztirārī*). This committee opened an account and asked the public to help with the financial situation. Money poured in (*Būltan khabarī*, issue 2, 27 Ordibehesht 2537).

At this point, the University Council launched a new and unprecedented resistance struggle. In the 1978–1979 academic year, it suggested accepting 700 students for the undergraduate program despite the regime's decision to close the school (*Rūzigār sharīf* 2016, 182). The acceptance of new students was advertised in a daily newspaper. The government commanded that all copies of the newspaper be confiscated, then published its own forged announcement claiming that the previous advertisement was false. The University Council resisted and published the original advertisement twice more, on August 8 and August 18, letting the public know that the university still planned to accept new undergraduates—regardless of what the government may have said about the matter.

Under these circumstances, the University Council could not conduct any kind of normal entrance exam,[38] so they had to quickly come up with an alternative recruitment procedure to acquire new students for the upcoming year. Ali Akbar Sayf Kurdi (faculty member) said that in May 1978, the faculty published 20,000 copies of an announcement that informed the public that the university would be accepting new students and, with the help of students, distributed these throughout the country. A blank application form accompanied each of these announcements. Prospective students were asked to fill out the application form and send it in. Approximately 14,000 applications were received in 1978, and the university accepted 480 students through this method. This alternative recruitment procedure was conducted in direct defiance of the will of the government (*Rūzigār sharīf* 2016, 92).[39]

On September 2, 1978, a group of professors at AMUT announced that members of the board of trustees and chancellor should be selected by the faculty—not appointed by the state. On September 6, 1978, the new government officially announced that it had accepted the demands of the faculty members, meaning the Tehran campus of AMUT would continue to operate. The government also agreed to the demand to allow the Isfahan campus to become its own independent university, with its own chancellor (*Rūzigār sharīf* 2016, 185). Moreover, the government also agreed to give the University Council the authority and responsibility for selecting its chancellor and making all its own major decisions regarding the administration of the university.

ACTS OF DEFIANCE AND DISSENT AT AMUT DURING THE IRANIAN REVOLUTION

The condition of Iran's prisons increasingly became the subject of global attention in the mid- to late 1970s, in the context of growing anti-Shah demonstrations both inside and outside the country, resulting in several instances of prison visits by the Red Cross. On October 25, 1978, approximately 1,500 political prisoners were released from various Iranian prisons. This action was carried out by the regime as part of a move to improve human rights conditions under such internal and external pressures.[40] On the same day that these political prisoners were released from various Iranian prisons, and while two universities in Tehran were shut down for the third consecutive week, *The Gazette* (October 26, 1978) reported that an "estimated 3,000 students" demonstrated at the AMUT campus. The students shouted slogans against the Shah and in praise of Ayatollah Ruhollah Khomeini, the chief opposition leader who was then living in Paris. The students were described as carrying photographs of "youths killed in recent street clashes between soldiers and demonstrators, and portraits of Muslim extremist guerrillas killed in action against security units" (*Gazette* October 26, 1978).

On November 6, 1978, the government shut down universities in Tehran, and they remained closed until January 1979. This decision came two days after the televised major student protest in November 1978. In December 1978, approximately seventy professors participated in a sit-in strike against the entry of University Guards onto the university campus (*Rūzigār sharīf* 2016, 190).[41] In January 1979, the University Council announced that the university would reopen in three days—without coordinating with the government or seeking state permission to do so (*Rūzigār sharīf* 2016, 191).

On February 10, 1979, one day prior to the complete overthrow of the Shah regime, SAVAK documented two demonstrations. One was a speech given in front of a crowd of 1,000 at AMUT by Abdolhassan Bani Sadr (1933–2021), an Iranian political dissident who had close ties to Khomeini. On the same day, a crowd of 3,000 gathered at the University of Tehran to observe the anniversary of the Siahkal incident.[42]

In this chapter, we have provided a detailed account of student activism at AMUT during the pivotal years directly preceding the 1979 revolution, examining the ideologies and the complex, and often contested, political, educational, and social demands of the student movement. Next, we turn to examine how these political events, conflicts, and contestations shaped cultural norms and practices at the university. The student experience will be the subject of the following two chapters. We begin by centering the experiences of female students and by scrutinizing the gender relations that prevailed in the revolutionary culture of AMUT.

5

GENDER RELATIONS AT AMUT

Following the state murder of Kurdish-Iranian woman Mahsa Amini, and the ensuing Woman, Life, Freedom (*Zan, Zindigī, Āzādī*) movement, once again the second-class citizenship of women under the Islamic Republic came into sharp relief. In the recent events related to the Israeli war on Gaza and the corresponding conflict between Iran and Israel, the Islamic Republic has ramped up arrests, harassment, and detainment of women who disobey the "moral" codes pertaining to how women should behave and dress in public spaces.[1] Given the primacy of gender justice in current protest movements in Iran and around the world, in this chapter we ask: What was the experience of female students at Aryamehr University of Technology (AMUT)? How were questions of gender taken up (or not) by the broader revolutionary culture of the university? How did these women navigate the dual pressures of state policies on "proper womanhood" and the prevailing leftist culture associated with student activists?

In the summer of 1968, Hengameh, a recent female high school graduate, was walking quickly through the campus of Polytechnic University,[2] located in the center of Tehran, on her way to view the results of her entrance exam. She was wearing a white shirt neatly tucked into a black skirt that hit her knees and a pair of black shoes with tiny egg-shaped heels. Her long hair was kept in a ponytail. This was the attire she chose on occasions that she wished to feel chic and professional. While we were interviewing her over the phone in 2021, talking to her from Chicago while she was at her home in Toronto, she shared with us her memory of the specific moment she found out she was admitted to AMUT. She was enroute to view the results of her entrance exam to Polytechnic University, when she encountered several male students who told her not to bother going all the way up to the office unless her name was "Elaheh." "Dismayed by the news," she recounts, "I learned only one female was admitted that year to the Polytechnic University, and her name was not Hengameh." However, it all turned out to be for the best, for she received the news from her father that same day that she had been admitted to study chemical engineering at her dream school, AMUT. Upon registering her, the administration had told her father, "The university has great plans, and the opportunity is massive" (interview with Hengameh, 2021).

In her first week at AMUT, to her surprise, everyone knew she was a first-year student. When she asked how all the other students knew this, they responded that there were only a limited number of female students,[3] and as such, everyone knew she was new to the campus (interview with Hengameh, 2021). She elaborated further:

The year I went to university, there were only 13 girls admitted who were mostly from Hadaf [high school]. . . . Oh, and I was going up the hill towards the campus where some boys were standing outside taunting, saying "look at the freshmen" . . . I was expecting to see a big crowd of students—like in Tehran University—and was wondering how these boys knew us?! When I went in, I realized that there were only 8 girls at university before us. (interview with Hengameh, 2021)

In recent years, about 70 percent of engineering graduates in Iran have been female,[4] and public perceptions of who is qualified to study science and technology are less gendered compared to the US and European nations. However, in the 1960s and 1970s, it was comparatively less common for women to be in engineering. Female high school students had to convince their families that studying engineering and later working as an engineer, which, at that time, was always imagined to involve work inside a factory, were not in conflict with ideas of proper womanhood (interview with Hengameh, 2021; interview with Batol, 2022). Many families doubted the engineering job market and its appropriateness for women, as well as how their daughters would fit into the competitive environment of AMUT.

Despite the gradual increase in the admission of female students, engineering was, broadly speaking, considered a masculine discipline in the 1960s and early 1970s in Iran, and those few female students remained a small minority of less than 10 percent at the AMUT campus throughout those years.

Even so, this small cohort of AMUT women perceived themselves to be expanding the career possibilities for women. In our research, we had the opportunity to speak to five women who attended AMUT in the years preceding the 1979 revolution. In our interview with Maryam, who entered as a freshman in 1972, she described majoring in industrial engineering because the name of the major included the term "industry" or "*san'at*": "Why did I choose my my major? well honestly, purely because it was connected to industry, and I was keen to join industry and become part of Iran's industrialization" (interview with Maryam, 2021). Similar to their male counterparts, female students saw within their engineering studies a path to join the societal push toward industrializing and modernizing the nation. However, for women in particular, the prominence attached to being an engineer in Iran was also about subverting the inferiority of their position in Iranian society. In this way, studying engineering at a prestigious university like AMUT was doubly significant.

CAMPUS CLIMATE AT AMUT FOR WOMEN

Unsurprisingly, female engineering students at AMUT faced many challenges, above and beyond those their male counterparts experienced. In addition to the academic demands and the complex political environment we discussed in the previous chapter, female students were also confronted with the sexist biases of their peers as well as of faculty and staff. In several of our interviewees, we heard stories about female students feeling judged when they asked questions in class. Another interviewee shared being mocked for not being able to carry some of the heavy laboratory items (interview with Maryam, 2021; interview with Maral, 2021). One of our male interviewees even described feeling unfairly evaluated on his written exams due to his unisex name, where presumably the professor carried implicit bias against female-appearing names (interview with Shiva Farahmand Rad, 2022).

On one level, unfortunately, it is not surprising that female students faced discriminatory behaviors and attitudes in the predominantly masculine context of AMUT. But how did revolutionary culture play into these dynamics? After all, as we described in detail in the previous chapter, AMUT was no ordinary engineering institution. The campus was ablaze with activism, dissent, and protest against the authoritarian dictatorship of the Shah. In our research, we were interested in understanding whether female students were engaged in the broader activism taking place at AMUT and what that experience was like for them as women. As one of our interviewees, Maryam, described:

Girls were also [politically] active. Of course, proportionately less, but we almost had 10% female students in each activity. For example, in the theater group where I was, and performed a play by Brecht, we had girls. In fact, girls were more inclined to be active, specifically those at the dorms, who had come from towns outside of Tehran and were experiencing more of a freedom in Tehran. There was more potential, and they could take better advantage of their opportunities to fulfill their dreams. (interview with Maryam, 2021)

Even the non–politically active students still intermingled with various activists and underground groups while participating in extracurricular activities on the campus, such as mountain climbing, acting in plays, and painting (interview with Maryam, 2021). Many of those who attended AMUT in the 1960s and 1970s participated in collective strikes, such as abstaining from taking final exams, including those who did not necessarily consider themselves politically active. Maral, an industrial engineering major, shared that while she was not directly involved in political groups during her student years, she couldn't "escape" activism on campus:

The atmosphere was extremely political. Debates, gatherings, cultural activities such as music groups, climbing groups etc., were all political and there was no escape! It could easily turn a non-political person to a political one. If someone wasn't politically active, it was perceived negatively. Even though we were only 18 when we entered the university. So, everything was influenced by the political atmosphere. (interview with Maral, 2021)

This environment affected student lifestyles, their social values, and even their refashioning of gender expectations. As another example, two years into her studies at AMUT, Hengameh began wearing her hair short. She wore button-down shirts with trousers. As she explains:

I used to wear trousers and "Chinese" shirts at that time. . . . It was similar to men's shirt, and we used to wear it over our trousers. . . . . . and as my dad was a factory owner, students would joke about me saying that I'd become a *chirīk* [as I was wearing casual clothes even though I was from a wealthy family] . . . that was a style, mostly wore by *chirīks* . . . I think it would've been too bold if I wore anything else, like a skirt. Or if I wore any makeup. (interview with Hengameh, 2021)

The female students began adopting behaviors considered masculine, and thus the performance of "female masculinity" became essential for female students to claim a revolutionary identity.[5] Some of these female students who displayed female masculinity were serious activists, while for others, their political engagement was limited to simply exhibiting their political sympathy with the opposition. However, in all cases, female students displayed their sympathies and allegiances through a performance of female masculinity. As Hengameh explains in the interview, the fashion she had adopted, such as wearing "men's shirts," helped her identify and be identified as a leftist (interview with Hengameh, 2021).

The revolutionary culture at AMUT, or what our interviewees referred to as the *mashy-i chirīkī* (i.e., the manners, social ideals, and worldview of OIPFG), led to the formation of an environment characterized by revolutionary masculinism, which, in short, is the valorization of certain modes of rugged masculinity and the repression of femininity. As mentioned earlier, the revolutionary culture (*mashy-i chirīkī*) became predominant on the campus of AMUT. For some students, the presence and gradual increase of women at AMUT became fodder for political conspiracies.

CONSPIRACIES OF FEMALE STUDENT PRESENCE IN IRAN'S REVOLUTIONARY CAMPUS

Due to the increasing predominance of *mashy-i chirīkī* on the campus between 1966 and the 1979 revolution, some of the male activists, particularly the more radicalized students, grew suspicious of the slowly increasing presence of female students (interview with Hassan Ghazimoradi, 2021). A conspiracy theory began to take hold that the state was sending female students to distract radical male activists from the revolutionary struggle—with the goal of eventually depoliticizing the campus. According to this theory, the presence of female students could lead to the possibility of romance on campus, which was considered a threat to the revolutionary movement. It is also important to see the interconnectedness of political activism by female students and their position as engineering students at AMUT. The architects of this conspiracy trafficked in the misogynistic beliefs pertaining to female students' diminished aptitude and competence in technical disciplines like engineering. Provided this belief, the increased presence of women on campus was attributable not to their academic merits but to the state's convert actions to infiltrate AMUT's campus with female students to presumably distract and seduce male students.

When asked about the presence and growth in the number of female students on campus, one of our interviewees, Behdad, mentioned being able to visibly notice the increase in large lecture halls. He mentioned being able to count on one hand the number of women in his first year but then noticing an observable increase each subsequent year. He explained how he thought about this then:

I admit I thought this was motivated by a political agenda. Increase the number of girls in engineering, as a way to distract the male students. And go after girls, romance, so and forth and such. This caused the decline of the engineering department at Sharif. [after a long pause, thinking] . . . If a girl was beautiful or attractive . . . [Radical students] would object to this. Investigate and scrutinize this . . . If a woman had makeup on for example. (interview with Behdad, 2021)

From this perspective, the normative political activist was male, and women were seen not only as incompatible with but even hostile to the revolutionary cause. Ultimately, the notion that a woman would simply choose to study at AMUT for intellectual purposes, or for proximity to political action, similar to male students, came to be viewed with deep skepticism by elements of the student movement. A particular set of masculine attributes had coalesced around the gender of the ideal activist at AMUT, particularly in the years preceding the 1979 revolution.

Industrial engineering student Maryam explained how, following the lead of more advanced students, she scarcely communicated with the opposite sex during the mountain climbing trips, as it was believed such encounters could be perceived as a potential for romance, which might serve as a distraction from the revolution. She described their mountain climbing activities:

The mountain climbing group was one of a kind! They would talk very seriously and wouldn't speak to girls. They had this serious (fighter) nature. [. . .]They didn't use to take us to two or three-day activities at first, but we insisted and managed to convince them to do it. Girls and boys would sleep together in the tents, although following very strict *"chirīkī"* moral rules. (interview with Maryam, 2021)

Hengameh, female alumni from AMUT, also mentioned, "It was very common amongst the hiking leftist groups to call people 'anti-revolutionary' if they started dating anyone. It used to be very extreme!" (interview with Hengameh, 2021). One of our male interviewees, Behdad, also explains, "It went to extreme positions such as 'why would a girl cross their legs?' or 'wear a short skirt?' This shows that they used to believe anything other than studying, that could encourage romantic relationships, would come in the way of the revolution. They also used to see that as part of the regime's plan to derail the revolution" (interview with Behdad, 2021).

THE TRAGIC LOVE AFFAIR OF EDNA SABET

The love affair of one of the famous female Marxist *chirīk* activists from AMUT, Edna Sabet, ended tragically. Some people speculate that her lover, Abdollah Panjehshahi, a fellow Marxist *chirīk* activist, was killed by his own comrades for his romance with Edna in their underground cell. In her 2021 presentation to the American Academy in Berlin titled, "When Love Was Forbidden: Sex and Intimacy in Iran's Revolutionary Generation," historian Naghmeh Sohrabi considers the murder of Panjeshahi by his comrades an important event for studying the revolutionary left subculture prior to the 1979 revolution (Sohrabi 2021). The event has also received attention from AMUT alumni such as Mohsen Seirafi Nejad, author of a recent book on the murder of Panjeshahi (Seirafi Nejad n.d.). Shiva Farahmand Rad has also engaged with the story in his review of the memoir of Marxist *chirīk* activist Maryam Satvat. As Farahmnad Rad explains, "Satvat's narratives shed light on the seeding of two love stories in the dry lands of underground cells where love was forbidden. One was between Edna Sabet and Abdollah Panjehshahi, and the other was between Satvat and Nima. The former led to the killing of Panjeshahi and subsequent Edna Sabet's mental breakdown, and the latter remained unarticulated in the name of the revolution" (Farahmand Rad 2019). The entanglement of the history of AMUT student activism with Panjeshahi's murder, as well as the living conditions of the couple in the underground cells, reveals the complex manner in which even the private lives of students were influenced, sometimes tragically, by the revolutionary culture at AMUT.

As we alluded to earlier, the sexual politics of the left at AMUT developed in response to the fear of campus depoliticization and eventual feminization of the opposition. In the context of Iran, OIPFG chose armed struggle against a state fully militarized. The influence of *mashy-i chirīkīc*, with its effect on valorizing revolutionary masculinism on the campus of AMUT, led to the adoption of more masculine behavior at the expense of the repression of certain other behaviors traditionally perceived as feminine. This was due to the equating of masculinity with power and femininity with weakness. We argue the general misogyny of the wider society became internalized by the activists—along with certain presumptions about masculinity having an automatic and natural correlation with the male body.

Nothing contradicts this misconception better than the fact that most of the male political activists at AMUT *also* had to work hard to meet the criteria of a proper revolutionary. While the presence of women in the movement was questioned due to them having female bodies, this did not mean the gender presentation of all men automatically conformed with these standards. The ideal activist had to engage in excessively strenuous mountain climbing, often while carrying a backpack full of heavy rocks with only a few dates for sustenance. The ideal activist also had to be someone who could deftly escape the authorities without having a safe place to sleep for days, or even weeks, at a time. He needed to be a person who would never surrender under torture, nor betray sensitive information about their comrades, even if threatened with death. He needed to abstain from romance and intimacy, and he had to love the working class, known as *khalq,* above and beyond his own personal interests (Pari n.d.). Given that AMUT was one of the most politicized campuses in Iran in the 1970s, these attributes transformed into wider social values and ideals and became essential in the construction of gender relations and conventions on a much larger scale.

CLAIMING A REVOLUTIONARY IDENTITY THROUGH PERFORMANCE OF FEMALE MASCULINITY

Our archival research revealed one particularly emblematic essay authored by a female student activist named Pari (pseudonym). In the essay, she delineates the ways in which women understood the role of masculinity in their activism as well as on AMUT's campus. The essay was published in the second issue of OIPFG's *Pāyām-i dānishjū* (*"A Message to the Students"*), and was titled "The Role of Females in Political and Union Activism." The preface to the essay, not written by Pari herself, begins by describing in detail a conflict between guerrilla fighters and SAVAK, culminating in Pari committing suicide by detonating a grenade after she ran out of cartridges when engaged in armed conflict with the police. Given that SAVAK had not learned her identity, the text does not reveal her real name and calls her by her pseudonym, Pari. The essay was published posthumously.

Pari's essay begins by declaring that female students have become active in political and union activism, and some have even become guerrilla fighters. She provides several examples, such as the strikes at the University of Tabriz, where female students accompanied their male counterparts in attacking guards with bricks, and at the University of Mashhad, where they had joined in physical conflicts with the guards. Through both these struggles, "they proved the idea that female activists are weak to be false" (Pari n.d.).

She provides other examples as well. For instance, Pari describes the case of a female student attacking a guard with her books to help secure the release of a male student from custody, which resulted in her own arrest. She described a particularly revealing incident at AMUT, when a large group of female students went to the chancellor, Seyyed Hossein Nasr, to protest the misogynistic remarks of the guard forces on the campus. As Pari tells it, horrified when confronted with such a crowd of female students, Nasr first tried to avoid them as a group by talking to just a single representative. However, the female students resisted, and Nasr finally submitted to their demands. According to this essay, Nasr had confessed he was afraid the group was ready to physically attack him (Pari n.d.).[6]

Through dramatic narratives of female activists' direct physical confrontations with police, Pari's infamous essay destabilizes the natural equivalence that is often assumed to exist between the male body and masculinity. The conclusion she drew was that female students are just as capable of political activism as men are, and that such behaviors constitute their identity as revolutionaries. Against the backdrop of revolutionary masculinism, female students performed female masculinity to display their equality with their male counterparts. And further, by performing female masculinity, female activists transformed their personal aesthetic into a trend on campus. As a result, other female students who were not politically active also adopted some performances of female masculinity, such as in their clothing choices and social values. They did this to express their sympathy with the activists and their opposition to the government.

Interestingly, the environment of revolutionary masculinism resulted in both the wide-spread rejection of femininity as weakness as well as the adoption of expressions of masculinity by female students. As an example, many female students who had sympathy with various factions of the left joined largely male contingents in physically demanding early morning hikes in the Alborz Mountains. Mountain climbing in Alborz was certainly not a leisurely break from studybut rather was intended to prepare student activists for an imminent revolutionary conflict with militarized forces. As Pari wrote, by joining these expeditions, "female students proved the backward idea of weakness of the female body to be false and baseless" (Pari n.d.). One of our interviewees, Maedeh, explained that after participating in mountain climbing with militant Marxists and attending other such extracurricular activities, she was no longer intimidated to ask questions in her classes. As she recounts,

Let me tell you something about being a women in engineering classes at AMUT. It was very intimidating. Any time I was going to ask a question my heart would beat so fast, and my voice would shake . . . it was really uncomfortable to ask a question with a shaky voice. But I had a real transformation . . . I remember a sense of confidence I developed from doing political things like mountain climbing. My understanding of my body and my confidence of mind changed. I could more comfortably ask questions without my voice shaking. (interview with Maedeh, 2021)

We highlight here how the adoption of female masculinity opened new intellectual subjectivities for female students. It not only reconfigured the gendering of the spaces, it also changed the way female students related to the broader political culture but also within the academic environment.

Pari's essay also discusses the participation of female students in political activism and how it allowed them to learn strategies for organization. For example, she describes how AMUT female students from outside of Tehran created a collective to demand their own dormitory, and following that victory, they established a library in the dormitory. The students effectively established a "home" for female students who arrived at AMUT from the outer provinces who could not afford to rent rooms in Tehran. They also democratized knowledge dissemination among financially disadvantaged students at the female dormitory. By challenging the gender regulations and being involved in political activism, activist female students reconfigured the campus to address class inequities and allow for the formation of new living spaces and social relations at the university.

One of the greatest casualties of this environment of revolutionary masculinism was pleasure and leisure. The tendency to attack concerts on campus was not limited to Islamists but also to members of OIPFG. One of the interviewees recalled a few instances of Islamist students attacking concerts on the campus and breaking instruments throughout their years at AMUT (interview with Behdad). Pari's article also gives us a nuanced understanding of how such praxis could be encouraged by an affiliate of OIPFG, in which the distinction between the practices of the leftists and Islamists becomes blurred:

In Iran's universities, there are students, misled by their false consciousness or dependency on the Pahlavi's state, who deceive students to purportedly participate in art forums such as theater and music, while in fact they are corrupting them with the ultimate purpose of them being depoliticized. The dirty role of the state in encouraging such endeavors is evident. (Pari n.d.)

The evidence the author offers for this is the timing of concerts and plays coinciding with student strikes or protests against the killing of the guerilla fighters by SAVAK. "In 1971–1972, when the student Khalil Tabatabaei was killed by SAVAK under torture, the music group of AMUT organized a concert to distract students from Tabatabaei's martyrdom. However, the defiant students attended the concert but attacked the musicians and singers with stones" (Pari n.d.).

The participation in the concert was criticized as being a tool to facilitate the normalization of government violence against student activists. Furthermore, pleasure and leisure were assumed to be the means by which the campus was depoliticized, and the state acted to disrupt or prevent political commitments. While some factions of Islamist students objected to concerts because they sincerely felt such entertainment was contrary to their religious beliefs, others viewed the concerts as impediments to revolution and political consciousness. In practice, however, it can be difficult to make this distinction. As Maedeh explains her observations of the campus,

Even though there was more of a leftist inclination at the university, the approach toward female students, or the relationship between them, was more of a religious kind. I mean, for example, those who looked more dressed up or chic, would be pointed out or warned. And the belief was that the regime is allowing girls to the university so the boys lose focus on their fight against the regime. [. . .] Unlike today, that the students come from top high schools, in those days students were mainly from more traditional families with old-fashioned mindsets. (interview with Maedeh, 2021)

However, another primary victim of revolutionary masculinism was a particular mode of femininity (equated with a lack of political responsibility and awareness). The adoption of masculinity as a rejection of pleasure-based femininity on the AMUT campus resonates with the gender critiques by the most well-known twentieth-century Iranian intellectual, Jalal Al-e Ahmad (Khanlarzadeh 2020 8). Al-e Ahmad's critiques of *qarbzadigī* (or West-Stricken-ness; 1962) advocated for the adoption of masculinity by women while also emphasizing that women must hold on to their role in safeguarding traditions. This principle was also discussed by Khanlarzadeh:

Al-e Ahmad's ideal woman was allowed to perform masculinity as long as she conformed to traditional values of femininity (such as modesty and preservation of traditions) and rejected modern leisure femininity (which was equated with qarbzadigī). This is similar to the gender ideology of Iranian Marxist guerrilla fighters in the 1970s, as Haideh Moghissi explains, "To break old barriers, women had only one option. They must behave like men and repudiate their femininity." (2020)

Despite OIPFG not allocating the role of safeguarding of traditions to women, the repudiation of certain modes of femininity and pleasure was similar between Al-e Ahmad and guerrilla fighters. Similar to OIPGF, Ali Shariati, the prominent twentieth-century Iranian intellectual and arguably one of the ideologues of the 1979 revolution, is not an advocate of women safeguarding traditions. However, he equates pleasure and consumerism with a lack of political consciousness:

In Shari'ati's view, women cannot flourish to their full potential as long as they live under the traditional cultural norms. However, he is equally critical of the modern replacements, i.e., being consumeristic, entertaining, and market-oriented. He argues that obtaining political consciousness and social responsibility instead of following old ways or new modern consumerism is the path to women's freedom. The equation of modern pleasures with superficiality and a lack of political consciousness is a limitation in Shari'ati's political thought, a view he shared with his contemporaneous Marxist thinkers. (Khanlarzadeh 2020, 16)

A shared conviction between OIPFG, Al-e Ahmad, and Shariati was that women were more prone to consumerism. As a result, they advocated that women eschew the pleasures of consumerism as the path toward acquiring political consciousness and raising the level of awareness in society. In hindsight, we can see how that theoretical orientation of intellectuals such as Al-e Ahmad and Shariati shaped certain practices and values of student activists at AMUT. For instance, besides the timing of the concerts, Pari suggests that the state utilized the engagement of female students with fashion and art to corrupt and depoliticize the student body and, in so doing, propagated harmful pleasure-based cultural values, as opposed to revolutionary ideals, among the students. She says:

Now the responsibility of the students in general, and female students in particular is clear. . . . The female students, in order to dismantle the obstacle set by the regime against the revolution, must raise awareness regarding the state's propaganda, and specifically target female students who, through fashion and such other quotidian activities, exhibit their susceptibility to corrupt capitalist cultural values. (Pari n.d.)

Similar to Al-e Ahmad's claim that women must take responsibility for their role in safeguarding Iranian and Islamic traditions, Pari argues that women should safeguard revolutionary ideals.

An important question remains unaddressed: Were the female activists at AMUT concerned with gender justice as part of their broader revolutionary struggle? Many scholars of the 1979 revolution have argued the left focused narrowly on capitalism as the primary cause of the exploitation of both men and women (Shahidian 1994; Moghissi 1994). By extension, a politics that centered gender was seen as a distraction from the primary source of oppression. As Iranian sociologist Haideh Moghissi has said, OIPFG believed:

The liberation of women depended on the restructuring of society. This would only materialize through the struggle of working-class men and women against the one and only enemy, that is, capitalism. In this interpretation of women's oppression, women had no specific identification apart from the interests of the working classes. (Moghissi 1994, 119)

Many activists maintained that only upon the success of a socialist revolution would both men and women be liberated. As such, any ostensibly narrow gender-focused politics were considered a distraction from the greater struggle for the liberation of Iranian people. Moghissi argues that in the broader Iranian society in the 1970s, "Women's interests were intertwined with the general goals of the anti-imperialist movement. Hence, women simply had to wait until all exploited and oppressed groups and classes were freed from the bonds of imperialist and capitalist relations" (1994, 118)

Iranian feminist scholar Manijeh Nasrabadi also engaged with this question regarding the female student activists of the Iranian Student Association (ISA) living in the US in the 1960s and 1970s. She finds that the desire among women activists for liberation from male domination stood against women's self-sacrificial affective investment in unity between men and women within the ISA, and in the broader success of the revolution. Nasrabadi maintains, "Women in the ISA were deeply invested in human liberation, including their own. At the same time, they were also affectively attached to notions of self-sacrifice, an indelible part of the experience of patriarchal social relations" (2014, 131). Thus, following Nasrabadi, the affective investments among ISA women (in unity with men and success of the revolution) were self-sacrificial, as these required them to downplay or altogether relinquish their gender-based liberatory aspirations. Furthermore, feminism was often considered a Western product or, more specifically, an intellectual-political characteristic of White American and European women. This held true among some other marginalized communities in the West as well. For instance, in the context of the US during the same time frame, Tracye Matthewis argues that the "whiteness" of feminism was often identified and critiqued: "Although early proponents of the WLM [Women's Liberation Movement] professed to encompass the issues, needs, and demands of all women, its initial definition of the term feminism, and its strategies, ideology, tactics, and membership, were dominated by white middle-class women" (2020, 237). Thus, in these cases, gender-based liberation was not considered the most crucial agenda of marginalized communities in their fights for racial and class justice in the 1960s and 1970s.

Similarly, the female activists of AMUT ideologically believed they were protecting themselves from the distractions of gender-specific concerns. Instead, they focused on dismantling capitalism in their anti-imperialist revolutionary movement. As we've described, the female students were imagined as *nūkhbah* agents of modernization of society, a subject position formerly perceived to be strictly male territory. This designation allowed female students to assume a more equal social position with their male counterparts and changed the gender norms with which their families disciplined and regulated their daily life practices.

As Maedeh recalled, she was allowed to stay out late in the evening for extracurricular activities (such as attending the theater or going mountain climbing) while her older sister was not allowed to do so (interview with Maedeh, 2021). She explained this difference was due to the recognition she had received from her family for being an engineering student at AMUT. This recognition can be attributed to the prestige gained by playing an active role in facilitating modernization and advancement of nation. And as their position with respect to the nation had changed, the gender rules governing their family life were also transformed.

Furthermore, the identification of female students with *khalq* allowed them to shift their perception of their class positioning and the manners in which they performed their class in their clothing choices and their social and political activities. As mentioned earlier, it became trendy for female students to keep their hair short, abstain from makeup, and wear button-down shirts. Hengameh, who studied chemical engineering at AMUT, and whose father owned a factory, explained that particular fashion choices allowed her to distance herself from her privileged class position and display her identification with *khalq*:

My dad's factory was in Karaj highway . . . as my dad was a factory owner, students would joke about me saying that I'd become a radical militant. . . . You know what, I think it was a way to distinguish oneself [through fashion]. I don't

want to say it was in fashion, but it was a way for you to show your mindset. . . .
I have a younger sister who was involved in all of these [political activities] from
high school. She used to tell me . . . She would call me a bourgeois for dressing
the way I used to before then. So, eventually my style changed to just a shirt and
a pair of trousers at some point. (interview with Henhameh, 2021)

In retrospect, the fashion choices of Hengameh and her sister might seem
performative given their access to resources and their upper-class privileges.
However, this allowed them to critique the prevailing society from a sub-
jective position different from the one they had inherited from their par-
ents. Further, this shift in class behavior also allowed the female students
to be creative and playful with their gender performances and participate
in the cultural practices and norms prevalent among their peers at AMUT.

Their stance against the militarized state also shifted how male students
performed gender and masculinity. They adopted certain behaviors and
manners, such as wearing plain, simple, modest clothing and abstaining
from consumerism and pleasure, to indicate their identification with the
khalq. This identification, which was shared among both male and female
students, resulted in the formation of new social relations outside of familial bond and dedicated to social improvement for the most oppressed sections of Iranian society.

A MOVEMENT CENTERED AROUND SACRIFICE

A search to form alternative communities for the purpose of imagining
and materializing social justice was not limited to the campus of AMUT.
Prominent Iranian Marxist intellectual Samad Behrangi, a hero in the
minds of the Iranian left and particularly among the OIPFG, wrote in his
stories about the formation of alternative communities consisting of toys,
various animals, and friends, all in search of liberation from oppressive
social-political relations and class domination (Khanlarzadeh 2021). The
student activists under the influence of *chirīkī* subculture shared these
goals described in Behrangi's stories and literary perspective. As discussed,
despite the alternative communities these spaces engendered, they were
still restrictive of intimate relations and pleasure.

The life goal of these activists, men and women, was the overthrow of
the Shah's regime. In their everyday manners and their adoption of certain

social values, they identified with guerrilla fighters. The AMUT students in particular were exposed to the struggle of the guerrilla fighters. Almost all the interviewees mentioned the story of a mathematician *chirīk* who was killed by SAVAK. One interviewee lamented, "He was a math genius, he was in the third year of electrical engineering, always smiled and seemed very gentle, asked the most interesting questions in the classes, I never knew he was a chirīk, until one day he disappeared and never appeared again, and later we learned he was killed by the SAVAK in a street conflict" (interview with Reza, 2021).

The guerrilla fighters sacrificed their self-interests in the name of an improved future for the *khalq*, and consequently, the activists valorized such sacrifices as well. Many of the female activists referenced their sacrifices performed in the 1970s to demand recognition of the role they played in the success of the revolution. For instance, in her article on the occasion of International Women's Day on March 8, 1979, Malakeh Mohammadi stated:

We oftentimes heard that a female activist in an underground collective house resisted the SAVAK forces to the last bullet so that she would create the opportunity for her comrades to escape, she sacrificed her life to free her comrades. They often put themselves in danger to free their comrades. When the regime shot them they shouted for the emancipation of the people, and the struggle of female students grew with time where they fought side by side the students of all universities for democracy and independence of universities and against imperialism. (1979, 47)

Based on such sacrifices, these same women later demanded their equal citizenship rights in postrevolution Iran. In the prevailing political culture, two main models existed for joining the political struggle and sacrificing for the success of the revolution: One was by taking political action, and the other one was through revolutionary speech. These models come from the third Shi'a Imam Hossein, who was martyred in the battle of Karbala, and his sister Zaynab, who gave a speech and raised awareness regarding her brother's revolution in the same battle. Shariati, the ideologue of the 1979 revolution, theorized their struggle and turned them into archetypes of resistance in the seventies. As explained by Khanlarzadeh:

After Hossein's martyrdom in Karbala on the Day of Ashura during the reign of the second Umayyad caliph Yazid, his sister Zaynab gave a speech to articulate the injustice that was brought to her family. Hers was not only a eulogy, it was a testimony to injustice; Zaynab thus symbolizes the historical truth, and the utterance of the word against injustice, through which the social awareness of society will be raised. In other words, Hossein represents the revolutionary act, and Zaynab represents the revolutionary speech. (2020, 14–15)

The notion of sacrifice was central to the philosophies and corresponding behaviors adopted by the revolutionaries and those who performed their political sympathies with them. It was under the influence of this political culture that female AMUT students either took action or performed their solidarity. In response to the loss of their classmates, music concerts and indulgences in romance and fashion became socially forbidden.

Performances of masculinity are not unique to AMUT students or to the Iranian political context. Significant historical and global parallels similarly have shown that when marginalized groups stand up against an establishment's militarized forces, the performances of excessive masculinity are seen through practices and social behaviors designed to shore up power and courage within targeted groups to sustain their resistance in the face of relentless persecution and terror. An example from the US context delineates this concept further. Carol Anderson explained that the decision by the Black Panthers to carry guns in public in response to police aggression was to demonstrate their courage as well as their knowledge of the law:

That lack of fear, that willingness to beat the police at their own game of aggression, that knowledge of the law and how it could be wielded to mystify and stymie the police, led Bobby Seale to define Huey P. Newton as "the baddest motherfucker in the world." (2021, 173)

The carrying of guns by Black Panthers was a direct response to the masculinist culture of the US police and militarism. Assata Shakur explains that during her time with the Black Panthers in the 1960s and 1970s, many of the female Panthers had to adopt a form of machismo to survive there. She explains she could not "say 'well listen brother, I think that . . . we should do this and this.' In order to be listened to," she recalls, "you had to just say, 'look mothafucka,' you know" (quoted in Matthews 2020, 243). In other words, women activists had to enact certain gender performances to be taken seriously. Shakur continues, "You had to develop this whole arrogant kind of macho style in order to be heard" (quoted in Matthews 2020, 243).

Revolutionary masculinism is in no way limited to the Iranian 1979 revolution or the Black Panthers. John Dollimore describes how a particular interpretation of manhood has often been thought to be essential in popular imaginings of resistance. He maintains, "There is an analogy here with working-class culture, black and white, where masculinity has also been a source of resistance, but once again at the expense of ratifying a larger exploitative framework for men as well as women" (2020, 34). Social movements concerned with class and racial justice oftentimes become oppressive in terms of gender and sexuality within the movements themselves. While the performance of excessive masculinity allows certain people to claim power and demand recognition, it can cause gender-related issues for women activists as well as male activists within the group who struggle—or refuse—to adopt a particular mode of masculinity.

Despite the predominance of revolutionary masculinism in both cases of leftist activism on the campus of AMUT and within the Black Panthers, aspects of the sexual politics at AMUT stand in stark contrast to the Black Panthers ,who encouraged "socialist fucking" among their members (Matthews 2020, 247). In the revolutionary culture of AMUT, romance and sexual relations specifically were considered a threat to the revolution. In one recollection, Maryam explained that students at AMUT condemned two classmates who had kissed each other in one of the classrooms:

They threw eggs at the girl and wanted to beat the boy. They even set a meeting about it! Some of the boys—as I heard later—set a meeting to discuss the major incident that happened at the university! It was decided at this (hidden) meeting that they would throw eggs at the girl when she is coming out of the lab and find the boy somewhere and beat him up! Imagine what a nightmare they created for that girl—what a lifetime trauma! What I want to say is that it's no surprise our revolution happened the way it did. The leftists were so extreme and had such traditional mannerisms. I was more of a moderate one, coming from a family of educators. I was inclined to the left since high school—more of a justice seeker kind—so I wasn't the target for those types of behaviors, but they existed. (interview with Maryam, 2021)

As we have shown, female students at AMUT adopted the performance of revolutionary masculinism to perform and practice their devotion to the social justice politics prevalent on their campus. Through the performance of female masculinity, they claimed the legitimacy of populating spaces formerly assumed to "naturally" belong to men. Thus, in practice, the female

student activists destabilized gender segregation through their engagement with extracurricular activities including mountain climbing, protesting against the state, and confronting the police and SAVAK. Throughout
these activities, they challenged the gendering of masculine spaces and the
perception that political activities were exclusively the prerogative of male
students, as well as the natural linkage imagined between the male body
and masculinity. The presence of female students at AMUT was initially
perceived critically by male students as a conspiratorial attempt by the
state to depoliticize the campus. However, by adopting the performance
of female masculinity and thereby demonstrating their belonging to and
sympathies with the revolutionary struggle, many female students successfully challenged the claims of these conspiracy theories. As Maryam noted,
by going mountain climbing with her male peers, she was both claiming
her belongingness to the revolution and, simultaneously, making herself
feel more empowered in the classroom.

Like their male counterparts, female engineering students at AMUT
participated in strikes and other political actions and thereby challenged
the gendered respectability expected from women by their families and
by the broader society. Once norms of respectability were challenged, new
horizons of gender possibility emerged, albeit still limited by the equating
of femininity with weakness and pleasure-seeking activities with a lack of
political consciousness. The ultimate challenge to the institution of family
and gendered respectability was the possibility of female activists joining
underground cells. Several AMUT female students from middle-class backgrounds did join militant groups and lived in underground cells. In these
cells, alternative collectivities were formed outside of familial bonds, and
this new form of home and family were shaped based on shared political
goals and affiliations.

Many female students of AMUT who joined the underground political
parties before the 1979 revolution continued their political activities after
the revolution and, thus, were often subjected to state violence a second
time. In her 2019 memoir, Maryam Satvat narrates the story of an AMUT
alumni named Zahra Zolfaghari, who was imprisoned both before and
after the 1979 revolution. Zolfaghari was a mechanical engineering student
who enrolled at AMUT in 1972. After her release from jail in postrevolution Iran, she developed a mental condition in which she imagined that

she was still living an underground life before the revolution and fighting against the Shah. At other times, she imagined herself to be in the years immediately after the revolution, actively fighting against the postrevolution establishment.

Our analysis of the experiences of AMUT female students has been informed by historians like Naghmeh Sohrabi, who has argued that a methodological shift is necessary for the historiographies of the 1979 revolution to better include the many women who were active in the movement. To do so, she argues, we need to move from the study of revolution-as-intellectual-history to the study of revolution-as-action (Sohrabi 2022). This pivot will bring women back into our narratives of the 1979 revolution. While women are less present in texts of prominent intellectuals of the time, they shaped revolutionary spaces without which the revolution would not have been possible. Sohrabi says, "Reading revolutions in this way [as action as opposed to texts of intellectuals] reveals a wealth of information that not only is crucial to the writing of social history but also allows us to move away, or at least question, the theorizing tendency that revolutions tend to spark in their scholars" (2022, 549). A powerful example of this mode of scholarship can be seen in the work of Iranian feminist scholar Manijeh Nasrabadi.

In her paper, "Women Can Do Anything Men Can Do: Gender and the Affects of Solidarity in the US Iranian Student Movement, 1961–1979," Nasrabadi investigates how the female student activists of the ISA living in the US in the 1960s and 1970s expressed their affective attachment to the revolution through what she calls the manner of *gender sameness*. Nasrabadi defines *gender sameness* as particular gender manners through which the women of ISA transcended their gender differences with men:

Gender sameness, as an ideology of women's liberation, rested upon the idea that by transcending gender difference and abandoning the degraded category of the "feminine," women could also transcend gender-based oppression. Sexual oppression was left untheorized and unchallenged, as was the gender binary itself. Iranian women participated in the regulation of their gender performance and sexuality because of their own desires for equality. (2014, p 132)

In the context of the specific cultural and gender politics at AMUT, the idea of gender sameness helps illuminate the convergence of social norms related to clothing and behavior expected of all revolutionary students.

However, to arrive at "sameness," it is important to note that men were not "becoming" like women, and sameness was achieved only through the efforts of women who had to transform their practices and behaviors. At the same time, men were also expected to adopt particularly narrow modes of masculinity and abstain from comfort and mundane pleasures. This led to an oppressive subculture where students whose gender manners were perceived as soft and feminine had a difficult time being accepted as serious activists.

The two main aspects of this particular mode of female masculinity were its repudiation of femininity and its aim of interpellating the students as revolutionary subjects. Consider the similarity here to the gender politics during the Chinese Cultural Revolution. The Chinese scholar Zhuying Li tells us, "Female masculinity means women should internalize masculine values and acquire power through engaging in masculinities, especially the hegemonic masculinity; and fully or mostly rejecting femininities in the meantime" (2014). While the performance of female masculinity does not necessarily mean a misogynistic rejection of femininity, both in the context of the Chinese Cultural Revolution as well as within the sections of the Iranian movement that led to the 1979 revolution, we see a rejection of femininity for its supposed association with weakness and lack of political consciousness.

Finally, it is important to consider the varying performances of female masculinity in Iran during the years leading up to the 1979 revolution. As Khanlarzadeh has explained, "There was a range of female masculinities in Iran of the 1960s and 1970s, performed by gender transgressive artists as well as advocated by several gender ideologies in their rejection of femininity" (2022, 8). From the 1950s until the 1979 revolution, in the lowbrow art scene of southern Tehran, for instance, female artists performed their particular adoption of female masculinity, which was tied to their off-stage gender identity, and thereby exerted their presence in urban spaces mostly populated by underclassmen. "Their performance of masculinity, in other words, was convincing to audiences due to their off-stage lifestyles, gender mannerisms, and stories of their 'manly' behavior in public" (Khanlarzadeh 2022, 6). The performance of gender in both the Iranian and Chinese revolutions did not reference women's quotidian gender identity, as was the case for female artists of southern Tehran. Instead, their gender expressions, both in China and Iran, became performances that indicated their sympathies with the revolution.

In the context of Iran, the performances of gender identity among female students referenced their transformations as revolutionaries and their belonging to a social group that refused to conform to the conventions constituted by the establishment and its normative discourses. To summarize, even though the female students of AMUT performed a certain style of masculinity, they were not identified as masculine women. Instead, they were known as revolutionary women.

6

POLITICAL IDENTITIES, LEARNING, AND SOCIAL EXPERIENCES OF AMUT STUDENTS

In this final chapter, we return to the main theme of our research: politicized engineering students in Iran's most elite university. We've discussed in depth the historical and sociopolitical conditions and context of AMUT, as well as practices and ideologies of dominant revolutionary culture. Now we turn to examine AMUT from the vantage point of the student experience. Drawing primarily on our oral histories, we begin by describing AMUT students' identities in conceptual and ethical terms. How did AMUT students think about themselves and their identities? How did their political commitments shape their views of engineering as a discipline, and what was the *experience* of being a politicized engineering student? Said differently, what did they care about, and why? The second half of this chapter explores these questions through a case study of one of our interviewees, Hassan Ghazimoradi.

THE ETHICS OF ENGINEERING AS LINKED TO MODERNIZATION POLITICS

In earlier chapters we've discussed modernization as the key philosophical and political priority attached to the founding of AMUT. Given the complex and fraught ethics of modernization as a political project, specifically one that has been linked to the politics of Westernization and

Western imperialism, we were keen on understanding how politicized AMUT students understood themselves with respect to these dynamics. A close analysis of our oral histories revealed the answer to this question to be exceedingly elusive and complex. On the one hand, as we discussed in previous chapters, AMUT students not only embraced modernization but saw themselves as agents of Iran's modernization. We heard time and again from our interviewees how their status as engineering students placed them in the great esteem of Iranian society, who viewed them as the worthy elites of the nation. In many ways, AMUT students basked in this admiration and took on a great sense of pride and responsibility toward their country. They were viewed, and viewed themselves, with a great sense of moral and political authority.

Interestingly, though, at the same time, AMUT students' political identities were rooted in a deep and unwavering identification with the oppressed in Iran, or the *khalq*. On the surface this appeared to us as contradictory ethics. How did their politics encompass identification with the most oppressed in Iranian society and, simultaneously, an embrace of the Shah's Western-endorsed politics of modernization? Early in our research, we paused to consider whether the students' political commitments were largely performative or immaterial. But as we read in chapter 4, AMUT students sacrificed mightily during their student years, some with their lives. Unquestionably, we found AMUT students' political commitments were neither superficial nor performative. Perhaps their most defining attribute was the sense of solidarity and collective responsibility they had to one another. While complex, heterogenous, and contradictory at times, their relational politics transcended mere philosophical alignment and materialized in their practices, decisions, the nature of their collectives, and, critically, their sacrifices, which often included putting their own lives and bodies on the line.

One of our female interviewees described it like this: "I remember we used to run, and the guards—villains—used to strike us in the legs from behind with their batons. My leg had become hard as a stone from repeated bruisings!" Another female student, Marya, shared this memory: "They would violently cancel the classes, university guards would start chasing us and beating us up. I remember one time I fell while running and the guard managed to catch me, struck me a few times but didn't ultimately manage to arrest me as the students gathered,and he had escaped." The real danger the students were in was widely recognized, including by their own professors who often acted as coconspirators:

The professors used to shelter the students in their rooms to save them from being arrested. I remember there was a professor—who was believed to be Baha'i—in the computer department. He helped gather the students and took those wounded to the hospital. And it's interesting that he would give fake names and addresses to anyone asking about his students. He was very protective.

In addition to the corporeal risks and peril associated with confrontations with police and guards, students also expressed their leftist politics through other manners. This was true particularly for AMUT students who were sympathetic but not directly linked to activist groups like the PMOI or OIPFG. One of our interviewees described how he unenrolled from AMUT for a period to work as a construction worker to better understand the predicament of the working class of Iranian society. Another interviewee described how volunteering as a high school teacher in Tehran's lower-income neighborhoods provided him with an opportunity to understand the political experiences and daily struggles of Tehran's urban youth. AMUT students' identification with the oppressed in Iran, expressed in multiple manners, was as foundational to their identities as their embrace of modernization as a political project. If AMUT students viewed their role as modernizers of Iran as inseparable from the struggle for political development and social justice for the oppressed, it raises the related question of how they understood the role of technology in Iranian society. Specifically, how were the ethical and cultural dimensions of technology understood? And how did AMUT students' political commitments shape their understanding of engineering as a discipline?

ETHICAL AND EPISTEMOLOGICAL POSITIONS ON TECHNOLOGY

As we described in the introduction, this study was informed, and in part motivated, by philosophical and cultural perspectives on science and technology, as well as critical scholarship in learning sciences and STEM education. Across this literature, common themes include questioning claims to rationality and objectivity, as well as critiques of how dominant Western values within science and technology disciplines have marginalized non-Western cultures and perspectives. Given the revolutionary political culture at AMUT discussed in depth in prior chapters, as researchers we anticipated AMUT students' views on engineering and technology to resonate with these critical perspectives. They didn't.

In the oral histories we conducted, by and large, students as well as faculty of AMUT expressed admiration and respect for the authority of Western science and rejected Islamic approaches to STEM. Many of our interviewees held fundamental epistemological positions on science and technology that align most closely with what some might describe as hierarchical and narrow views of science and technology. In fact, many of our interviewees reasoned that the inherent rationalism and objectivity of STEM disciplines are precisely what qualifies them to analyze and provide commentary on social and political problems. A former AMUT professor explained to us this way: "A mechanical or electrical engineer could nicely design their machines and earn good money. This makes them believe that politics work like science. That you can design and expect it to work!" He continued, albeit critically, discussing the thinking of Mehdi Bazargan, who was the head of the first engineering department at the University of Tehran and also the first prime minister of Iran after the 1979 revolution. "Bazargan knew a bit of thermodynamics and used to think that the whole world is run by it! In his book *Thermodynamics of Humans: Love and Worship* he is trying to explain human emotions through thermodynamics! It's impossible." (interview with Professor Ahmadiyan [name pseudonymized], a former professor at AMUT, 2022)

In assessing the epistemological views of AMUT students and faculty, it is important to recognize the history of philosophy of science and technology in Iran, and specifically the intellectual history of a school of thought known as nativism. In the context of Iran, nativism was a philosophical and cultural movement that emerged in the early twentieth century in response to Western imperialism and colonialism. Nativist thought in Iran was often intertwined with epistemics of Islamic philosophy and advocated for an embrace of traditionalism, Indigenous knowledge, and cultural perspectives in scientific inquiry (Boroujerdi 1996; Mirsepassi 2017).[1] Thus, just as AMUT students' politicization must be understood inside the particularities of Iran's intellectual and cultural-political context, similarly, their philosophical views regarding science and engineering were directly related to a rebellion against the orthodoxies of Islamic ideology and nativist philosophy.[2]

In responding to our question regarding the tendency of engineering students to pursue politics, one of the interviewees reveals this connection clearly:

Those who study fundamental science or engineering are involved with subjects such as math and physics which teach you logic and how to scientifically analyze the world. When approached with sociopolitical matters, science provides this critical thinking capability, and that's one of the reasons why religious fundamentalists do not approve of science. So, approaching fundamental religious questions with a scientific mindset, can result in doubting and questioning. (interview with Arash, 2020)

Another professor spoke bluntly about what he perceived as the problem of bringing religion and science together: "It hurts me to see ideological views influencing science. Whether it's Marxism or Islam. I believe science doesn't have philosophy. Physics doesn't follow philosophy and those involved in it were only looking for ways to discover the world" (interview with Ahmaiyan [name pseudonymized], 2022). Across our interviews, we learned that for AMUT students, the rationalism and objectivity of science stood in stark opposition to the dogmas of Islamic ideology. And in this way, affiliations with Western science carried a rebellious and political appeal a political appeal to students as they negotiated their academic and political identities during their university years at AMUT.

Our discussions about the philosophy of science also revealed the complexity of how AMUT students thought about the ethics and values of STEM disciplines more broadly. In one discussion with an interviewee, the conversation turned to the unforgiving moral assessments of Western science by many Iranian intellectuals of the twentieth century. Our respondent shared his own criticisms of those thinkers:

We can't stop the advancement of science for the fear of misusing the technology .. morality has such complicated meaning to it. What do we mean by morality? For instance, Iranian intellectuals say a moral approach to science is to not use it for military and wars, right? An example is the immoral US bombardment of Nagasaki, in Japan. Don't you think it would have been immoral of other countries to not prevent Germans to get access to the nuclear bomb? (interview with Ahmadiyan [name pseudonymized], 2022)

In the previous quote, our interviewee is referring to the tradition of Western criticism in the intellectual history of Iran, embodied in the scholarly works of writers like Chancellor Nasr, Jalal Al-e- Ahmad, Ahmad Kasravi, and Fakhr al-Din Shadiman. We note that ethical perspectives of AMUT students and faculty were complex and sophisticated and transcended simplistic rejections or blind allegiances to Western paradigms. Importantly, while rejecting nativist positions, we also saw articulations of ethics that were culturally and politically specific to the context of Iran. Echoing intellectual giant Ahmad Kasravi, who wrote incisively about the corrupting role of militarism in science during World War II, the former professor discussed his position on ethics in this way:

Morality has such complicated meaning to it. What do we mean by morality? For instance, they say a moral approach to science is to not use it for military and wars, right? Don't you think that the scientists of Iran needed to do something when the country was being bombarded by Iraq during the war? Now, is it moral or immoral of me not to engage in military work, as a scientist? (interview with Ahmadiyan [name pseudonymized], former professor at AMUT, 2022)

We turn to explore these contradictions and complexities through the case of Hassan Ghazimoradi.

HASSAN GHAZIMORADI

We chose Hassan because not only did he clearly embody the attributes described earlier but he was one of the few participants who stayed in Iran after completing his studies. Significantly, his engagement with politics as well as engineering far outlived his university years. Although many of his peers were killed or imprisoned, or moved to the West and pursued careers in technology within industry or academia, Hassan remained in Iran and worked as an engineer for decades while simultaneously deepening his political engagement.[3] His physical location and ongoing cultural-intellectual engagements with the political condition of Iran affords a unique perspective that we wanted to uplift and honor. Furthermore, Hassan was a student during Nasr's term as chancellor, providing a unique opportunity to explore the meaning and implications of Nasr's politics at

AMUT. And last, while Iranians are known to be artful speakers and story-tellers, we found Hassan to be especially gifted in this regard.

Hassan attended AMUT from 1972 to 1977, which was two decades after the 1953 US–British-sponsored coup that toppled nationalist Prime Minister Mohammad Mosaddegh. By 1972, the year that Hassan entered AMUT, radical organizations were recruiting many of their members from among the university students. The university was also constantly under pressure from the Shah's intelligence service, and the behavior of students and professors was closely monitored daily (NLI, 230–035171). Hassan's attendance at AMUT coincided with the chancellorship of Dr. Seyyed Hossein Nasr. In 1973, when Hassan entered his second year, more than one-third of students (1,400 students out of 3,600) lost their right to enroll in the university because they had participated in such protests (Ganjīnah [AMUT Historical Document Center]). Finally, 1977, the year that Hassan graduated, marks the beginning of the nationwide popular anti-Shah demonstrations that resulted in the toppling of the Shah in 1979—one of the most important revolutions in the twentieth century.

Hassan Ghazimoradi was both a typical and an atypical engineering student during his years at AMUT. Like his peers, he was politically engaged and philosophically aligned with *mashy-i chirīkī*, the leftist-Marxist political subculture of the university. While he was an engineering student, Ghazimoradi was as serious about his politics as he was about his academics. For him, these were not separate spheres but rather deeply interconnected parts of his identity. Indeed, though initially a physics major, Hassan switched majors to engineering in his first year because he viewed engineering as offering a more direct path to gaining employment immediately after graduation and remaining closely involved with political issues:

I started enjoying physics when I was at the second year of high school. So, I decided to study physics and selected it as my first choice at the Sanati University and got accepted. I joined Sanati University as a physics student. However, I was introduced to political activities and student movements at the University. This meant that I decided it was no longer appropriate for me to just follow my dream of studying physics and not get involved with the challenges of the society . . . after the first year I decided to change my subject . . . I changed my major to mechanical engineering and graduated in it. (interview with Hassan Ghazimoradi, 2021)

Also like many of his peers, Hassan was from a lower-middle-class and moderately religious (Islamic) family. He was unequivocally critical of the Shah during his youth and college years. Early in the interview, he jokes with us about how he (and his peer group) would refer to the AMUT as just "University of Technology" as opposed to its official name starting with the Shah's epithet "Aryamehr":

I need to point out that we didn't use to call the University 'Aryamehr.' It was called The Sanati (industrial) University of Aryamehr, but we used to call it Sanati University. We never used to use that word [Aryamehr (Shah's title)] and it's obvious why! So, I will be referring to it as Sanati university in this interview. We are used to it as the students from those days. Anyone having certain political views in those days [against the regime] would also specifically avoid it. (interview with Hassan Ghazimoradi, 2021)

The omission (or rejection) of the use of "Aryamehr" is a subtle but significant linguistic rendering of the general anti-Shah sentiment that Hassan and his peers embraced. For Hassan, the experience of being an engineering student was inseparable from politics. As Hassan says, "Being an engineering student was not just about academics. It was about being political, being active, being engaging socially." During his AMUT years, along with his studies, Hassan taught mathematics in local high schools in working-class neighborhoods of southern Tehran, which he credits for giving him a deeper understanding of the political concerns of Iranian youth.

VIGNETTES INSPIRED BY HASSAN'S EXPERIENCES AT AMUT

Through Hassan's case, we have arrived at a more detailed understanding of an individual student's experiences with respect to the complex, layered, and often contradictory politics of AMUT. The stories contained in our interviews with Hassan were deeply rich and textured, and we became more and more interested in finding ways to reconstruct aspects of his experiences to best illustrate AMUT from a first-person account. In this section, we present three fictional vignettes based on Hassan's experience and inspired by the method of critical fabulation, an innovative artistic methodology for conducting historical inquiry.[4] As we present the case of Hassan Ghazimoradi, critical fabulation provides a way to artfully evoke the lived experience of being a politicized engineering student at AMUT. We extend Hartman's use of archive to include oral histories as an additional source of inspiration for the development of the vignettes. We interviewed Hassan on three separate occasions, amounting to nearly seven hours of interview data. The interviews were conducted in Persian, then translated into English and carefully analyzed for themes relevant to our research question. The fabulations are fictional storytelling elements that were carefully added to highlight the emotional and affective dimensions of Hassan's experiences while taking care to accurately reflect the essence and meaning of these experiences as perceived by Hassan.[5] All three vignettes are written in third person in present tense, incorporating archivally sourced ethnographic details that enliven the richness of Hassan's learning experiences in and out of the classroom, on and off campus.

VIGNETTE 1: "THE POLICE ARE OUTSIDE"

Hassan shows the guard his ID at the campus entrance as he approaches Mojtahedi Hall, named after Dr. Mohammed Mojtahedi, a founder and former chancellor of AMUT. As he walks in, he picks up his exam sheet and takes off his jacket before sitting down. The sweat on his palms reflects the storm of emotions occupying his mind, the adrenaline pumping through his body. Scanning the room, he finds solace in the faces of his peers who look scared but also proud. Today is exam day—a notoriously difficult exam to evaluate their knowledge of the topics covered in the first few months of linear algebra—orthogonal matrices, eigenvalues, linear transformations. But Hassan's nerves have nothing to do with these obtuse topics. In fact, he has admirably mastered them. Like many of his peers, the intellectual rigor of AMUT's curriculum, derived directly from its partnership with the Massachusetts Institute of Technology (MIT), is a source of pride and respect. Yet, their allegiances today are rooted elsewhere: Students are boycotting all final exams in an act of political solidarity with leftist groups opposing the Shah of Iran. Hassan places his winter scarf under his right hand to prevent the moisture from dampening the exam sheet. Out of habit, he begins writing a verse from the poet Hafiz in Persian calligraphy,

"To the profligate, world-consuming who hath abandoned the world's attachments what business with counsel-considering, The land's work is such that deliberation and reflection is necessary for it."

رند عالم سوز را با مصلحت بینی چه کار

کار مُلک است آنکه تقدیر و تامل بایدش

He feels deeply connected to the words as he ponders the possible consequences for his participation in the boycott—the university has rapidly militarized in the last few weeks. His friend Morteza, seated in the row ahead, slightly turns his head toward Hassan and ominously whispers, "The police are outside. They have sent a van. Wait outside at the bus station for me." Hassan's breath quickens, and he wonders to himself: What will happen if I am arrested? How will I explain this to my family? These thoughts are interrupted by a pang of guilt—his mind fixing on the image of two students from the electrical engineering department who were gunned down a few months earlier by government officials shortly after their detention. Suddenly, there are shouts outside and the rustling sounds of students packing their bags, shuffling with quiet urgency toward the exit. In his hurry, Hassan leaves behind his jacket and waits in the cold for Morteza at the bus station.

This scene illuminates the complex interplay between learning, identity, ideology, and politics that Hassan frequently encountered and negotiated as an engineering student (Philip et al. 2018). Hassan's relationship with engineering was informed by his view that engineering as a discipline carries significant implications for the advancement of Iran as a country—economically, politically, and militaristically. Given Iran's sociopolitical status as a developing country, engineers were indispensable to the technological capabilities of the nation, and in this sense, studying engineering was automatically a politicized decision. In the story, these values are reflected in Hassan's intellectual stance toward his linear algebra class. He was studious and prideful that he was excelling in the rigorous MIT-affiliated curriculum that he was offered at AMUT. We note that despite his leftist politics, he was uncritical of the endorsement AMUT received from a prominent American institution. In this sense, his politics were not tied to a challenge of Eurocentrism or Western knowledge production. This is particularly notable given the specific vision for Islamizing and Iranianizing the university advanced by Nasr and reflected in the Center for Humanities established by Nasr. In our interviews with him, when asked about the epistemological possibilities for scientific knowledge production in non-Western contexts such as Iran, Hassan reasserted positivist perspectives—asserting the neutrality and objectivity of scientific knowledge. Indeed, for Hassan, the proximity of the curriculum of AMUT to elite Western knowledge was a source of *'iftikhār* (اف تخار), a pride and eagerness to learn the disciplinary topics central to a modern and global engineering education. Even so, for Hassan, his academic identity existed in complex and dynamic relation to his leftist and political identity. In participating in a student-coordinated boycott of the linear algebra exam, he engages in a highly politicized act of disobedience and, we note, does so at a great risk to himself. Many of his peers experienced severe persecution, including death in many cases, for political activities during this period in Iranian history. Thus, we understand Hassan's participation in this boycott as tied to a deeply rooted sense of collective responsibility he has to his peers (Ma et al. 2016), and a political identity that coexists with but takes precedence over his academic identity. The boycott scene demonstrates how even inside the engineering classroom—in this case, a linear algebra class where learning is focused on core disciplinary topics—students at AMUT were negotiating far more than the content of the curriculum. The disciplinary learning that occurs for Hassan and his peers is cooccurring with learning about self, solidarity, and sacrifice.

VIGNETTE 2: THE TRIAL

It was one of those sunny, glorious days in Tehran. Hassan walks through campus on his way to the central library—trying to discern the meaning behind the averted glances, the careful whispers of students gathered in small groups. Finally, he pieces together the story. He learns that the evening prior, two students had kissed each other in one of the classrooms of the chemical engineering building. As he walks toward the library, Hassan hears students accusatorily muttering in hushed tones, muḥḥākimah bāyad bishan *(محاکمه باید بشن) (they must be put on trial). In the lobby of the library, he sees his friend Bijan, a student activist, and inevitably the conversation turns to the impending trial. Bijan explains the crime as a moral judgment. Hassan listens while Bijan makes his case but is unmoved by Bijan's argument. As Hassan and Bijan say goodbye, Bijan notices a philosophy book in Hassan's hands. Displeased with what he perceives as another kind of distraction, Bijan scolds Hassan for his interest in reading at this critical juncture in history. Now is the time for action, not theory. Hassan feels a sense of fatigue. He retreats upstairs to the sanctuary of the library and closes the door to the outside world. The librarian asks him if he has heard of the campus uproar.*

Without revealing his admiration for the kissing students, Hassan simply nods his head in affirmation. As he walks away, he smiles slyly and whispers to himself, damishūn garm *(دمشون گرم) (good for them).*

This scene highlights the tensions and contradictions present in the practices and ideological stances of AMUT's revolutionary student movement, and significantly, how they present a complex intellectual and moral terrain students were forced to navigate. In our interview with him, Hassans admits he at times he fell victim to the conspiracy theories we described in the previous chapter related to the presence of women on the AMUT campus. He describes these prevailing conspiracies as unwavering evidence of undemocratic and regressive elements in the political culture of student activists. The vignette further elucidates elements among certain aspects of the teachings of the left, which Hassan finds uncouth, namely, the tendency toward anti-intellectualism, in which even reading philosophy is seen as a distraction from the revolution. Hassan, a young man barely twenty years of age, resents these stipulations, and we contend his refusals are simultaneously signs of his identify formation as well as powerful moments of sensemaking and learning. In being provoked to grapple with the moral and intellectual demands presented in this scene, we witness Hassan learning how to think independently and stitch together a complex and nuanced political identity. He becomes a person who simultaneously stands in solidarity with the left while simultaneously rejecting *some* elements of the left's discourse and cultural practices. Hassan's rejection of a particular cultural practice of the left signifies his ability to think critically and independently. While inspired by the social justice aspirations of the left, for Hassan, this did not translate into a blind adoption of or unquestioning admiration of its complex social practices and ideals. As a young man, Hassan vibrantly illustrates a carefully constructed political identity that exists simultaneously within and against the political culture of the left at AMUT.

VIGNETTE 3: "TULIPS ARE RISING FROM THE BLOOD OF OUR YOUTH"

Philosophy class just ended. Hassan is weaving through traffic, accompanied by Ostad (Professor) Khoei, a legend in Iranian literacy circles. Khoei works as a visiting professor in the new Center for Humanities on campus. He doesn't have a car and has taken up Hassan on his offer to provide him a ride to his desired destinations, usually to local maykadahs میکده, *Iranian taverns populated with writers, activists, and intellectuals in prerevolutionary Iran. Enthusiastic to spend time with Khoei, Hassan borrows his brother's 1975 Paykan on days that he has class with him. During the rides, they usually discuss poetry and philosophy. While in the car today, Hassan is playing a recently released cassette tape by the singer Mohammad Reza Shajarian:*

"tulips are rising from the blood of our youths, the cedars in the streets have become bent grieving the loss of our youths' cedar-like stature."

از خون جوانان وطن لاله دمیده

از ماتم سرو قدشان، سروها خمیده

The lyrics refer to a song composed at the time of Iran's Constitutional Revolution (1905–1911), mourning the losses of the revolution. The song's melody blends with the blaring of car horns and seductively mingles with the plumes of smoke emerging from the cigarette dangling at the edge of Khoei's lips. They arrive at their destination. Right before the notoriously lengthy departing greetings commonplace in Iranian social relations, Hassan finds the courage to ask his professor for a favor: "Aghayeh (Mr.) Khoei, I know you are busy, but I have a draft of a new fictional story I have written. Will you read it and tell me what you think? I would be eternally grateful." Khoei graciously agrees before stepping into the maykadeh میکده *and waving goodbye to Hassan.*

In contrasting to the first two vignettes, here Hassan is in the driver seat, literally and figuratively. Through his own agency, he has constructed an intellectual "third space" (Gutiérrez et al. 1999), leveraging the resources AMUT provided through the Center of Humanities (a renowned poet-philosopher) toward his personal learning and development. Here, by offering to be his personal chauffeur, Hassan has skillfully extended his student-professor relationship with Khoei, learning from him in the more relaxed context of friendship (Jackson et al. 2020), engaging in informal *guftigū* (dialogue) while weaving through Tehran's notoriously hostile traffic.

Learning and identity processes are intertwined here in several ways. First, there is the knowledge he is gaining from the intellectual discourse itself, from sharing space and time with a literary legend discussing the nuances of contemporary poetry and its complex relations to the cultural-political landscape of prerevolutionary Iran. For Hassan, again at merely twenty years old, these short trips with Khoei represent remarkably profound opportunities to extend his literary, philosophical, and political knowledge. Significantly, this learning is not passive. Hassan is also learning how to engage as a worthy interlocutor in what many might consider an intellectually intimidating encounter with a poet of Khoei's stature. His musical selection is subtle, purposeful, and illustrative of Hassan's situational awareness and agency. Shajarian's mournful melodies in the background convey a shared political sensibility and cultural appreciation. In these moves, Hassan is thus embracing an identity of an engineering student whose interests and talents extend well beyond scientific or technical matters. To engage in politics at AMUT entailed an inherent connection to the arts, history, intellectual pursuits, and philosophical ideas.

Across all the vignettes, Hassan was engaged in high-stakes ethical decision-making (e.g., in vignette 1, deciding to boycott in solidarity with leftists, or in vignette 2, refusing the anti-intellectualism or regressive sexual politics of particular elements of the left). The scenes simultaneously offer a snapshot of his political identity in a moment in time and capture the deep sensemaking that constitutes the process of Hassan's political formation. Taken together, the learning scenes also illuminate the complexity, contradictions, and tensions of politicization at AMUT. Hassan was learning to be a politicized engineering student, for him a process that entailed both alignment with and rejection of particular ideas and cultural practices of the left on the AMUT campus.

Though differentially visible across the scenes, and notwithstanding the contradictions, the presence of a radical and vibrant student movement at AMUT was unquestionably the most significant factor in Hassan's political development and learning. In our final interview with him, Hassan clarified that the politicization of students occurred despite the best efforts of the university to suppress dissent. Even the Center for Humanities, noted Hassan, while offering a rich array of cultural, philosophical, and artistic learning opportunities, was for many students ultimately linked with the cultural politics of the Shah (Matin-Asgari 2018) and, thus, was viewed with skepticism by many on the left. This was true despite the fact the

Center had invited many highly regarded thinkers and dissidents, such as Shamloo, who had been imprisoned for his poetry.

Ultimately, for Hassan, the roar of the radical student movement thrived despite the best efforts of the Shah, Nasr, and other university leaders. Furthermore, while he embraced the intellectual opportunities afforded by the Center of Humanities (as we see in vignette 3), he ultimately interprets the learning offerings in Islamic philosophy or Persian literature as thinly veiled tactics to distract and appease radical students. It comes as no surprise then, that he scoffs at the notion of an "Islamic chemistry" or a "Persian physics" and rather views serious engagement with Western science and technology as a pragmatic and serious strategy for Iran's social and political advancement (interview with Hassan Ghazimoradi, October 17, 2021).

Our discussion of AMUT students' experiences is consistent with how scholars have described the role of political identity in the life course. For example, the experiences of AMUT students like Hassan resonate with how social psychologists have described the development of political identity as an "outward-looking process in which [students] anticipate their lives as adults and struggle to understand who they are within a social and historical framework" (Yates and Youniss 1988, 495). They go on to argue, "As part of this effort, [students] reflect on the values, ideologies, and traditions of their communities and the possible roles they will undertake in adulthood" (Yates and Youniss 1988, 495). At AMUT, student identities were constructed in relation to the complex cultural and political context discussed in depth in previous chapters.

Furthermore, political identities at AMUT were intricately tied to the idea of intellectual elites (نخبه *nūkhbah*) and how this idea has been historically perceived in the context of a developing nation that is undergoing a centralized and state-sponsored industrialization effort. That is, AMUT engineering students were viewed (and viewed themselves) as not just intellectually talented but also obligated in a sense to provide moral and political leadership for the nation. And finally, we must consider the complex and sometimes contradictory logics that gave rise to the university. On the one hand, AMUT was founded by the Shah and controlled by the royal court with the explicit intent to "modernize" the country and *prevent* the kinds of student dissent commonplace in other Iranian universities.

On the other hand, AMUT was the site of a unique experiment in Iranianizing the university,[6] as well as the site of a formidable and highly radicalized student movement. All of these factors shaped the ideological and material contexts within which AMUT students made sense of themselves and who they were in relation to their nation. These complexities and contradictions materialized into the physical and symbolic subculture of the university, and how the politicized identities of engineering students emerged at AMUT and existed in complex relation to these configurations.

CONCLUSION

The AMUT students and faculty discussed in this book were temporally located between the 1953 coup and what would become the most significant popular revolution of the twentieth century. The significance of these historical bookends is inscribed spatially on the AMUT campus: The boulevard across from where these students sat in their science and engineering courses carried in its name—Eisenhower—the traces of Western imperialism. This referred to the president of the US (1953–1961) who, on August 19, 1953, approved the US–British coup against Iran's prime minister, Mohammad Mosaddegh, and his nationalization of the petroleum industry. After the revolution, the boulevard was renamed to Azadi (meaning freedom, آزادی)—a change quite emblematic of what the students had hoped to achieve.

In this book, we have analyzed the political experiences of students and faculty at AMUT in the years directly preceding the 1979 revolution. We examined the historical and political processes through which the new university took shape in Iran and the ways engineering students and faculty experienced this highly dynamic and complex period in Iranian history. Few other scholarly studies have theorized or empirically investigated how sensemaking about the values of STEM disciplines are dynamically connected to students' political identities. That is, whereas racial and gender identity have been conceptualized as a distinct phenomenon that dynamically interacts with STEM learning, students' political identities with respect to STEM learning has scarcely been foregrounded (see Morgan et al. 2020 for a recent exception). This includes conceptualizing people's ethical sensemaking about the politicized dimensions of the disciplines themselves as worthy of theory and study.

In this book, by taking up and extending theories of identity in STEM education, we have examined how learning environments are imbued, sometimes explicitly and other times in more subtle ways, with ethical meaning (Hess and McAvoy 2015), and how these meanings have implications for individual sensemaking around the values of a particular discipline, as well as sensemaking around one's desire to pursue learning in a particular discipline or domain. For instance, as alluded to earlier, in the US context, research has demonstrated that students often associate engineering and computing disciplines as embodying values antithetical to justice and human rights concerns (Garibay 2015). How, where, and through what processes do students learn these associations? Might these associations be disrupted or reframed? How do the values students attach to particular disciplinary domains have implications for their future patterns of engagement and achievement, for their selection of colleges and majors, and, ultimately, for their career and life pathways? These are open empirical questions with significant pedagogical and policy implications that come into focus when moral and political identity is taken up as a fundamental aspect of human activity and behavior. Lastly, and critically, how can research in diverse global and historical contexts challenge existing understandings of how identities are constructed in relation to disciplines such as engineering? These are all questions we hope this book has not only raised but also began to address.

AMUT AND IMPLICATIONS FOR STEM EDUCATION

The case of AMUT is an example of how global perspectives, particularly from the Global South, can deepen and complicate contemporary discussions on ethics, epistemology, and knowledge production in STEM fields. First, let us examine an important backdrop to these discussions. In the last several years, critical STEM education and learning sciences scholars have deepened the engagements of our field with questions of power, ethics, race, and justice in STEM education. For example, critiques of STEM as tied to militarism (Vossoughi and Vakil 2018), capitalism (Morales-Doyle and Gutstein 2019), coloniality (Bang et al. 2014) and anti-Blackness (Nxumalo and Ross 2019) have all challenged the epistemological and ontological foundations of STEM education research and practice: "Inclusion toward what ends? Whose knowledge? What counts as science? Who gets to decide?" These questions, in fact, were central in the selection of AMUT as the focus of our historical inquiry. As Iranian diasporic scholars, the case of AMUT presented a remarkable opportunity to contribute a Global South perspective to an expanding and exciting dialogue on alternative epistemologies in science and technology.

But, alas, in this regard, we did not learn what we expected to learn. As we discussed in the previous chapter, a key finding was that in the case of AMUT, questions tied to alternative (and specifically non-Western) scientific epistemologies did not resonate for revolutionary students in the ways we had expected. To contextualize this, let us recall the anti-imperialist politics of the late twentieth century that reverberated powerfully on the AMUT campus. Let us also recall that AMUT engineering students were directly influenced by, as well as influencing, freedom and liberatory movements in the US, Latin America, Africa, and China. Yet still, while the relationship to engineering of former students was inseparable from their deep investment in revolutionary politics, it was also strongly linked to a situated understanding of technology as critical for the advancement of Iranian industry, military capacities, and economic independence. What is noteworthy here is that these are precisely the associations that, in the Western context, often produce incentives for politically aware students to *disengage* with engineering disciplines (Garibay 2015).

This is instructive: The ethical implications of proximity of technology to structures and institutions of power are not static but instead highly dynamic, and, in some cases, inverted, depending on geopolitical factors and dynamics. In the West, engineering as a project of modernization and industrialization often can contradict core commitments of social justice and liberatory politics. However, in the context of Iran and other developing nations, where nation-building can be interwoven with questions of democracy, freedom, and social justice in complex and unexpected ways, we contend that engineering may very well require a different political calculus. As we have shown, modernization via engineering was precisely the Shah's political vision for the new university. Despite opposition to the Shah as a defining attribute of the revolutionary culture of the campus of AMUT, like the Shah, the centrality of engineering to the advancement of the nation still figured prominently into how students understood themselves. Notably, the desire to actively participate in the modernization process of Iran that we observed in our interviewees is powerfully mirrored in the quest for status that Iran as a nation has aspired to on the global stage.

In the context of a nation such as Iran, which was struggling to fight off foreign interests and assert itself in a contested and often hostile political environment, technology was seen as a tool of resistance, respect, and dignity. This is directly relevant to the epistemological questions as well. For many AMUT students, though rooted in opposition to a US-backed authoritarian regime, their politics did not readily translate to critiques of the university's MIT-affiliated curriculum or, more broadly, to cultural criticisms embodied by thinkers such as Nasr, centering on the spiritual and moral impoverishment of Western science and technology. We contend that sitting with these dissonances is the crucial work necessary to expand our conceptions of what counts as "critical" or "ethical" engagement with science and technology in diverse global contexts. We return again to the words of Professor Ahmadiyan:

Morality has such complicated meaning to it. What do we mean by morality? For instance, they say a moral approach to science is to not use it for military and wars, right? Don't you think that the scientists of Iran needed to do something when the country was being bombarded by Iraq during the war? You may not remember that time, but they dropped a bomb behind our home, and I remember 18 kids were killed. Now, is it moral or immoral of me not to engage in military work, as a scientist?[1]

Does knowledge and development in STEM for self-defense, as Ahmadiyan implies, cast conversations about ethics, militarism, and science in a different light? These questions do not have easy answers, but we feel they deserve careful consideration. In this regard, the case of AMUT demonstrates the benefit of engaging global contexts in thinking through challenging questions related to ethics and epistemology in STEM fields. The fields of education, science and technology studies, and learning sciences are experiencing momentous theoretical advance. Scholars have effectively scrutinized the false and restrictive boundaries of disciplinary learning, calling rather for notions of epistemic heterogeneity (Barajas-Lopéz and Bang 2018; Tan, Barton, and Benavides 2019), and multiple ways of knowing (Warren et al. 2020) to guide theories and pedagogies, particularly with respect to historically marginalized populations.

The fact that Iranian revolutionary students in the 1970s do not fit neatly into these narratives does not diminish this essential, ongoing work to broaden and deepen the epistemic and ontological possibilities for STEM learning. However, we assert that viewing *revolutionary Iranian students'* mode of political engagement with STEM as assimilationist or reductive would also be misguided. At AMUT, engineering students demonstrated a keen sensitivity to and sophisticated understanding of the relation between technology and politics in Iranian society.

NOTES ON POLITICAL ENGINEERS IN THE MIDDLE EAST

We would be remiss if we were to ignore the particular ways in which the political orientation of engineers has been of keen interest to the realm of national security, the military and intelligence community, as well as to many social and political scientists. This also includes domestic efforts to surveil Muslim communities, as well as recruiting academics to understand the specific mechanics of how people get radicalized in Islamic societies. As it turns out, one proposed answer to this question has been connected to the core of the engineering discipline itself. In the book *Engineers of Jihad: Curious Connection between Violent Extremism and Education* (Gambetta and Hertog 2018), social scientists Diego Gambetta and Steffen Hertog examine why a disproportionate number of Islamists and right-wing extremists have engineering degrees, as compared to left-wing radicals, who are more commonly groomed in the social sciences and humanities. They argue that the quest for social mobility along with an attraction to order and hierarchy inherent in the engineering discipline are key factors that explain the overrepresentation of radical Islamists. This is a provocative and complex claim that raises hard questions about inherent values and dispositions of disciplinary domains, but we urge caution against reaching any premature conclusions or making any facile connections between the discipline of engineering and political identities in the context of Muslim and Middle Eastern nations. Far too many complexities, uninvestigated subtleties, and unknown factors are still in operation here. In a review of the book published in the *European Sociological Review*, Lagrange (2017) states:

The prominent role of engineers in violent extremism and in jihadist groups in Muslim countries is nevertheless puzzling. Affinities between science and engineering with violent Islamism and right-wing extremism on the one hand and affinities between humanities and the left-wing violent extremism on the other remain to be better understood.

We take issue with the implied association between studying engineering and violence for particular populations in the Islamic world. Furthermore, the research we conducted for this book challenges the findings of such prior studies in important ways. First, obscured in superficial comparisons between Islamic radicalism and engineering is the stunningly rich history of contributions the Islamic world has made to the fields of science and technology (Al-Hassani 2012; Hogendijk and Sabra 2003). Second, as we have demonstrated in this book, the basic premise that engineering students in the Middle East have a propensity for right-wing Islamic politics, rather than left-wing social justice causes, is patently false in the case of Iran.

To be sure, Iranian engineers are far from a political monolith. There have been politicians, conservatives, mullahs, and imams with engineering backgrounds. Former conservative president Mahmoud Ahmadinejad earned a doctorate in civil engineering and was a practicing engineer before entering politics. Indeed, Mehdi Bazargan, founder of the Islamic Association of Engineers in Iran, was appointed by Ayatollah Khomeini as the Islamic Republic's first prime minister. Revolutionary philosopher Ali Shariati was among the occasional lecturers at the organization. There is also a separate organization named The Islamic Society of Engineers, which defines itself squarely as a political organization aligned with Iran's ruling clergy and committed to promoting the Islamic, scientific, and technical knowledge of the Muslim people in Iran while guarding against "foreign cultural agents" from the West. It was formed in 1988 at the end of the grueling Iran-Iraq war and exists today as a nonprofit governmental organization. However, it has also overwhelmingly been the case that engineering students have frequently been active participants in—if not leaders of—leftist, progressive movements with roots in Marxist as well as third-world internationalist politics. As such, engineering as a discipline in the Iranian context is closely associated with anti-establishment, progressive politics—not right-wing or conservative politics.

ARYAMEHR UNIVERSITY OF TECHNOLOGY BECOMES SHARIF AFTER 1979 REVOLUTION

Immediately after the 1979 revolution, AMUT was renamed Tehran University of Technology. However, soon after, the name of the university was changed once again to Sharif University of Technology, after Majid Sharif-Vaghefi, a former electrical engineering student at AMUT and leading member of the People's Mujahedin of Iran (PMOI), who was killed during an internal purge of non-Marxist members. Today, Sharif University remains the most prestigious university in Iran, boasting over 10,000 undergraduate and graduate students within fifteen main departments. It is still widely considered the Iranian MIT. Since the 1979 revolution, a great deal has transpired in Iran politically, all of which has deeply influenced university life and politics. The Islamic Republic's Cultural Revolution (1980–1983), the eight-year Iran-Iraq War (September 1980 to August 1988), the mass execution of political prisoners in late 1980s that resulted in the killing of several AMUT alumni, the chained murder of public intellectuals (1988–1998), the reform movement (1997–2005), the Green movement of 2009, and the recent uprisings of the Woman, Life, Freedom (*Zan, Zindigī, Āzādi*) movement. It is quite remarkable that through it all, Sharif University of Technology has not only remained open and operational but has persisted as a key site of demonstrations and political struggle. The numerous news accounts of the government's siege of Sharif University during 2022 are only the most recent manifestations of this enduring legacy.

However, we must ask, how has Sharif University of Technology fared with respect to the original aspirations of the Shah, Nasr, Mojtahedi, and Amin, among other university leaders who envisioned a bold and expansive education that would produce technically sophisticated and culturally aware engineers committed to the future of Iran. For Nasr, there is no question. The 1979 revolution destroyed the liberatory vision and possibility that AMUT had for the future of Iran. He explains this in his interview with Hossein Ziya'i:

Two or three years before the 1979 Revolution, more than fifty percent of Iran's industry was managed by AMUT students. So, if this was not stopped by the Revolution, Iran's most industrial sections could be run by AMUT engineers in 10 years, and thus, Iran could be independent in its industrial development.

Further, Nasr notes the deleterious impact of the revolution on faculty retention. In our interview with him, he discussed how the revolution stimulated a mass departure:

So, what happened was that, when the revolution took place in 1979, we had 450 PhD professors at Aryamehr university. This is not the people who were assistant, with masters, teacher etc. I mean professors. Of these, only 72 remained. The exact statistics.

Overall, our discussion with Nasr made abundantly clear his assessment that the quality of engineering education has been severely compromised over the last several decades. Notably, this is in alignment with one of the faculty interviews we conducted:

You see . . . After the revolution, the universities were forced to employ those preferred by the government. The only university that resisted against this decision and didn't drop their employment standards, was Sharif. On one occasion we were *told* that we have three new faculty in physics. We said great! but we won't give them any courses to teach as we can't trust the standards by which they've been chosen. They [government] eventually gave up, and we managed to keep our standards at the University.

When asked to comment on how changes to Sharif University since the revolution has shifted the landscape of scientists and engineers in country, he said:

On the surface, we have strong scientists and computer programmers who have done great things such as developing Snapp [Iranian equivalent of Uber] and Digikala [Iranian e-commerce company]. However, those in power don't have an innovative, long-term vision about science. The mindset is mainly to take advantage of science to solve the current challenges of the industry. They are not focused on helping to improve the science in the world but rather how to copy other countries to deal with their current issues. I believe, even in today's Iran, when we talk about 'research', we are more concerned with the prestige that comes with it rather that its actual benefits.

Importantly, this critique of science and technology, tethered narrowly to the needs of industry, is being levied against Sharif today echoes criticisms of AMUT before the 1979 revolution. As the former professor made clear, "Science holds a strong power in the world that unfortunately our politicians, heads of states and kings never understood. There are aspects to science that make it more powerful than politics."

It is quite interesting that the misdirected and unfulfilled potential of science and technology was also precisely the original raison d'être of AMUT. By any account, AMUT was a grand experiment in modernizing Iran through the delicate and contested endeavor of combining Iranian and Islamic culture and values with a Western scientific conceptual framework. However, although this may well have been a noble effort, it ultimately seems to have failed in its core mission due to a complex set of social and political factors. Even so, the scale of the ecological crisis in Iran, the undemocratic restrictions on the media, the internet, communication technologies, and the ongoing militarization of technology industries make it clear that science and technology in Iran after the revolution has definitely *not* reached its potential as a liberatory or democraticizing force. While we cannot lay the blame upon individual institutions such as Sharif University, given the ongoing repression and rampant corruption of the government and its institutions, the role of engineering as a force of good in Iranian society and politics remains doubtful—if not doomed.

SHARIF UNIVERSITY AND THE CONTINUING POLITICAL CONFLICT IN IRAN

In chapter 4, we discussed a famous sit-in strike at AMUT in 1977, which occurred on the brink of the revolution. At that time, the Shah was being hosted by President Carter in the US. Due to Carter's widely lauded stance on human rights, many expected that the regime would de-escalate matters during the Shah's visit. However, they were quite wrong. As noted in the chapter, that sit-in became a decisive moment in the student movement of that time. Given AMUT's history of activism, it was not surprising that during the 2022 student movement in Iran, one of the early surges of the protest movement on Iranian campuses occurred at Sharif University in October 2022. During a meeting held at Sharif on October 2, 2022,

Mohammad Ali Zolfigol, minister of science, research, and technology, did not mince words in his condemnation of students and faculty participating in the campus protests. As a warning, Zolfigol told the protestors, "You cannot say whatever you want without consequences." In an attempt to identify, trap, arrest, and terrorize student protesters, on this day, security forces, many of whom were undercover, besieged the Sharif campus (Fassihi, 2022). The residents of Tehran were quick to respond. As soon as news circulated on social media about the crackdown at the Sharif campus, as a show of solidarity with endangered students and faculty, hundreds of supporters rushed to the university. This event, known as the Siege of Sharif University, became a hugely influential moment depicting student resistance and popular support in face of state brutality.

As we were finalizing the initial draft of this manuscript, we approached the one-year anniversary of the state killing of Mahsa Amini (September 16, 2023). In the spirit of honoring and amplifying the voices of current student activists, we now turn briefly to hear their brave voices amid the chaos and despair that has gripped our homeland. The most recent issue of *Forough* (issue 19, August 2023), the journal of the Islamic Association of Students at Sharif University, is dedicated to several women's rights activists such as Homa Darabi, a university professor who self-immolated in 1994, as a protest against mandatory hijab, and Sepideh Reshnu, a poet-activist who openly protested state-imposed hijab. This issue features a graphical representation of the symbolic act of Vida Muvahhid, depicting her standing on a utility box, removing her scarf, and attaching it to a stick, which became the symbol of an antimandatory hijab movement known as the Women of the Enghelab Street. In this issue, several students from Sharif, as well as other universities in Iran, have reflected on various aspects of the political movement. The cover of this issue also bears the meaningful slogan: "We have crossed the threshold of fear." In this same issue, Kianush Babayan, a mechanical engineering student, has provided a compelling account of the current political atmosphere at Sharif University.

Perhaps some of the actions during the first half of 2022 were predicated on the analysis that, due to societal passivity and prevailing sociopolitical conditions, coupled with the two-year hiatus in on-campus activities, the university would lack the required dynamism to counter coercion and repression. However, the students' relative resistance during this time, including two significant protests against security measures at Tehran University of Science and Technology and Tabriz University of Medical Sciences, demonstrated that despite the disruptions caused by the pandemic and the somewhat passive state of society at the beginning of 2022, the university retained its capacity to resist. With the unfortunate passing of Mahsa Amini, this resistance materialized in action, as the university followed the streets' [movement] from day one of the protests. (Source: Babayan, *Forugh*, issue 19)

Tragically, as Babayan explains, the killing of Mahsa Amini served as a catalyst, as university students joined the street protests from the very onset. To be sure, the Islamic Republic government is actively and insidiously working to maintain control over students by implementing new regulations and adjusting the disciplinary code through its supervisory and executive branches within the university, particularly through the security and disciplinary committees. This concerted effort is aimed at diminishing the university's vitality and vibrancy. Moreover, the issuance of numerous suspensions and educational restrictions, the revocation of activity permits for student organizations, the termination of contracts with specific professors, and the imposition of additional financial burdens on students who participated in the autumn 2022 protests collectively highlight the government's pursuit of retribution against the institution. In his statement, Babayan argues that this retaliatory action is a direct consequence of the institution's dynamic and pivotal role during the autumn protests.

Finally, we would be remiss if we did not mention the experiences of AMUT students who were killed or imprisoned. These fallen and largely forgotten former Iranian engineering students from AMUT are the ghosts who haunt and animate the stories we shared throughout the book. We have labored to pay tribute to them, partially by subjecting the political-cultural context of their lives, dreams, ideas, and practices to careful social-scientific study.

We did not know what we would find, and what we did find was at times hopeful and inspirational, and other times painful and complicated. Ultimately, this was the story of how an elite university founded during the reign of the Shah by the Shah himself ironically contributed to his downfall, and persists in contemporary Iran as a training ground for elite Iranian scientific and technological talent as well as a site of ongoing dissent and resistance.

EPILOGUE

ECHOES OF DISSENT: EXPLORING THE LEGACY OF AMUT STUDENT MOVEMENT AND CONTEMPORARY REACTIONS

While conducting the research for this book over the past couple years, we have shared bits and pieces of it at various conferences, including Association for Iranian Studies (AIS) in Salamanca, Spain, the Middle East Studies Association (MESA) in Denver, Comparative and International Education Society (CIES) held virtually in Minneapolis, as well as a Brown Bag Series held by Middle East and North African (MENA) studies at Northwestern University in May 2022. We have also discussed the subject of the book extensively within our own networks of friends and family. Across these various communities, we have received quite the range of reactions and responses to our investigation and to its implications. The overall reactions and interest in the topics of this book have been undoubtedly heightened by the killing of Mahsa Amini and the Woman, Life, Freedom (*Zan, Zindigī, Āzādi*) movement. Most of the reactions to our project have been overwhelmingly supportive, but there were also some critical reactions to the project, particularly from members of the Iranian diaspora. The questions—sometimes explicitly stated, other times implied—challenged the premise of the project: *Why waste our time studying leftist Iranian students whose superficial and misguided politics contributed to the*

1979 revolution that delivered the lethal, tyrannical regime that stubbornly remains in power? Why even return to the political events that have caused such trauma in our community?

Several significant contemporary Iranian scholars, writers, and artists have explored the psychological dimensions of the collective trauma and mourning experienced by the Iranian diaspora in the decades since the 1979 revolution. For some, especially for those who may have once been quite critical of the Shah's regime, including leftists or those deeply affiliated with leftists, but who since have felt disillusioned or betrayed by the revolution, this mourning has sometimes translated into feelings of guilt. In other instances, sadly, there may be feelings of resentment toward the left for contributing to the ascendance of Khomeini and the current despotic regime. From this perspective, dedicating a study to the lives, thoughts, and experiences of these students seems futile or, perhaps even worse, a kind of glorification of leftist anti-Shah students. We respect these sentiments and desire to treat them with care in our final remarks. In *Teaching Community: A Pedagogy of Hope,* feminist scholar bell hooks writes,

It is my deep belief that in talking about the past, in understanding the things that have happened to us, we can heal and go forward. Some people believe that it is best to put the past behind you, to never speak about the events that have happened that have hurt or wounded us, and this is their way of coping—but coping is not healing. By confronting the past without shame we are free of its hold on us. (hooks, 2003, 119)

We regard the notion that AMUT's history is unworthy of close examination because of its role in bringing about the revolution as a form of coping by Iranians in the diaspora who have lost their homeland to religious extremists. But as hooks points out in the quote, coping is not healing. The research we undertook here has been equally informed by our scholarly interests as well as by our personal investment in more deeply understanding the causes, sources, and discourses of the 1979 revolution. Our project has sought to combine rigorous research with our personal desires to understand the trajectories of our own lives and those of our families. This is our way to not merely *cope* with the complicated and often tragic and melancholic realities of diasporic existence but also to heal and to move toward new futures.

At a fundamental level, it can be said that irrespective of what transpired after the revolution, our investigation has been anchored in a
respect for the aspirations and sacrifices of the revolutionary engineering students who boldly dreamed of a brighter, freer future for Iran, and
whose political and cultural identities were intricately woven together
with their identities as engineering students. As we have demonstrated
extensively throughout the book, this orientation has not prevented our
careful scrutiny, and even critiques, of student practices or ideas.

Intriguingly, in the oral histories, several interviewees were themselves
quite critical of their own earlier actions and ideas as they reconstructed
them with us in the interviews. For instance, as one of our interviewees,
Hengameh, remarked, "We may have been passionate, but we were stu-
pid! We were also very persistent but had no idea what we were doing!"
(interview with Hengameh, 2021). The same interviewee mentioned that
her father, who owned a factory, questioned whether the political circumstances were really as horrible as students imagined them to be. Hengameh said, "My father always teased that the university is giving you
students cheap kebabs, you have filled your stomachs and have become
violent" (interview with Hengameh, 2021). Echoing the sentiments of
her father, she joked, "I remember, for example, some protests about the
food quality, although it was very good!"

Alongside and interwoven with these light-hearted reflections were
also more serious moments of self-reflection and critique. Looking back
over four decades, one of the interviewees, Maryam, described a kind of
political naivete, even pretentiousness, that existed within the revolutionary culture:

In the mountain group they used to read very emotional poems. To be honest,
there wasn't much difference between a leftist like me and someone who wasn't
a leftist. Maybe some of us leftists had read a few more books. Some hadn't read
any. These are the adventures of the young age that can manifest in such ways.
We used to think we're doing such important things! sleeping in sleeping bags,
waking up early and conquering the mountains. (interview with Maryam, 2021)

To be clear, we do not understand these kinds of self-deprecations of former selves as evidence of the unseriousness of revolutionary students or
their efforts. Quite the opposite. We understand these critical reflections
as evidence of individuals, some forty to fifty years later, who remain

keenly interested in the significance, meanings, and consequences of who they were as people in an extraordinary era of their lives. Many of them had devoted themselves to a cause larger than themselves at significant personal risk, whatever we might now say about the outcomes of those efforts. And with the benefit that only multiple decades of thinking and reflection can provide, they have developed sharp—and sometimes even ungenerous—analyses of their own actions and behaviors. We regard this as a demonstration of wisdom and humility.

UNCHARTED HISTORIES OF AMUT AND PROSPECTS FOR FUTURE RESEARCH

Writing a single book would scarcely suffice to fully explore the extensive history of AMUT and its profound legacy, considering its multifaceted contributions not only to Iran but also to the global realm of engineering and STEM education, enriched by its highly esteemed professors and alumni. Even with our timeline's emphasis on tracing the institution's history from its inception to the revolution, myriad perspectives exist through which one can approach AMUT's story. Nevertheless, our primary focus in this book has been to comprehend specific facets of its history, guided by the questions outlined in the introduction. However, it's imperative to acknowledge the multitude of other narratives and histories that remain untold within our narrative. For instance, the dedication of numerous individuals, including the early professors who pioneered teaching at AMUT and significantly contributed to the development of various departments. Noteworthy examples include the pivotal role played by Dr. Firooz Partovi, a graduate of MIT, in shaping the Physics Department, or the contributions of exemplary figures like Morteza Anvari to the early development of the Center for Humanities, as well as the role of certain students such as Shiva Farahmand Rad in managing student-led music libraries. While these are just a few examples, many others exemplify themes within AMUT's history that warrant rigorous scholarship and research.

Furthermore, it's crucial to recognize the institutional ties between AMUT and the history and development of Alborz College, which originated as the American College of Tehran under the directorship of Presbyterian missionary Samuel L. Jordan before being nationalized in 1940. As noted in the book, not only was Mojtahidi, the first chancellor of AMUT, also the manager of Alborz, but he also leveraged his social and intellectual networks among Alborz alumni to recruit professors and staff. Additionally, the establishment of the new university was made possible by the dedication of numerous staff at Alborz who donated nontrivial portions of their salaries. These interconnected relationships necessitate a deeper exploration of the history of Alborz College.[1] Undoubtedly, there's much more to be said about the various ways these two institutions, particularly the legacy and educational culture of Alborz College, influenced the early development and identity of AMUT. These are areas in which we encourage further research.

IMPLICATIONS FOR ONGOING STUDENT POLITICAL ACTIVISM

As we are nearing the completion of this book, a new wave of radical student activism is sweeping across college campuses in the United States, including here at Northwestern University. The ongoing Israeli assault on Gaza has reignited fervent discussions and protests concerning Palestine, specifically addressing critiques of imperialism and the role of technology in warfare. Notably, students on campuses such as Northwestern, UCLA, MIT, Harvard, Columbia, New York University, and University of Texas, among many others, are not just protesting the war on Gaza but also fundamentally challenging US imperialism. In this regard, they have much in common with the revolutionary students of AMUT.

At an institutional level, the political context of AMUT, and the implications it had on student identity development, carries significant implications for current efforts to support and cultivate socially aware science and engineering students and faculty. Based on the experiences and responses of AMUT students, university leaders invested in cultivating the social agency and sense of responsibility of students would be wise to support conditions on college campuses in which student movements can exist and even thrive. STEM students on our campuses today, like those in Iran during the 1970s, and like engineering and science students during the civil rights era when universities were ablaze with student activism (Renehan 2007), will undoubtedly benefit from thinking and developing alongside healthy, vibrant student movements on the campuses where they study and live. Ultimately, the most significant factor in the politicization of AMUT engineering students was the massive popular uprising that swept across Iran, including within institutions like the Aryamehr University of Technology.

APPENDIX: ORAL HISTORY PROTOCOL

INTERVIEWING THE FORMER STUDENTS OF AMUT

People interviewees here were students at AMUT during our research period from the year 1966 to 1979.

Introduction

Hello. My name is. . . . The purpose we follow in this interview is to explore and develop a better understanding of pedagogical and research methods and the ways former students of AMUT experienced the cultural and political environment of the University. For this purpose, we are interviewing former students of AMUT. This interview will take up to two or three hours, and the questions center on your experiences during the time you were a student at AMUT, including matters related to education and the cultural and political environment of the University. Our purpose is to understand your experiences better, so there are no wrong or right answers. If you agree, we will record this interview. The purpose for recording this interview is to not lose any of the material that you share with us. However, this interview will remain confidential, and your name or characteristics that would reveal your identity will not be shared in future publications. Please study and sign the consent forms. If you have questions regarding these forms, please don't hesitate to ask us. (The forms will be given to interviewees and will be signed.)

INITIAL QUESTIONS

- Introduce yourself: name, birth date, and birthplace
- In what years were you a student at AMUT?
- What was your major? From what city did you come to AMUT?
- After you graduated, where did you get a job and in what field, or in what university did you continue your higher education?

QUESTIONS RELATED TO EXPERIENCES AS A STUDENT

We would like to learn more about your experiences during your student years at AMUT:

- Why did you choose your major?
- What do you think about the pedagogical methods at AMUT: the way courses were designed, the ways professors taught, the books that were chosen for students, and in what language those books were?
- From the way the courses were taught, the material of the courses, and the research that was conducted, did you feel AMUT was attempting to adapt science and technology, in any way, to Iranian and Islamic culture and history?
- How many female students were admitted to AMUT in your year, and how many of those were in your major? What is your observation of the gender policies and gender realities of the time at the university? Did women participate in political and social activism? Do you remember any particular example?
- What was the role of religion in the classroom, in social events, political activism, policies, and the overall environments of the university? How did you relate to the role religion was playing, at the time, in the formation of resistance to the government?
- At the time that you were a student, the White Revolution and the Cold War were in the political background. How did you experience these in the university?
- Do you think the university at the level of officials was formulating any specific political goal in the teachings and research of science and technology?

- If you have experienced studying or teaching in another university, how do you compare that other experience with your experience at AMUT?
- What do you think about the concept of *nūkhbah*? *Nūkhbah* acts like a social category in which students of AMUT are placed. How did you experience this? How does the concept of *nūkhbah* shape the characters of AMUT students?

QUESTIONS RELATED TO THE 1979 REVOLUTION

- Were you part of any social or political activities at the university? Did your activities begin while you were a university student, or from the time you were at high school? If yes, please explain.
- Did you work with any particular political group?
- Do you remember whether political and social discussions happened during the classes? If so, were there any consequences for the students or professors?
- In your opinion, what was the role of AMUT in the formation of the resistance movement that led to the 1979 revolution?
- Who were the well-known student or professor activists at the university at that time? Did you know them in person? What did you know about them and their activism?
- What do you think about student political activism? Do you think there was anything particular about the activism of students of AMUT, or it was no different from the activism of other students from other universities?
- Do you remember the Siahkal incident or any other significant political event of the time? What was the reaction in the university? How did you relate to the incident? Were you a student at the time?
- It is often a point of fascination for observers that students of engineering and sciences are very political in Iran. How do you make sense of this as a former student of science and technology?

INTERVIEW WITH THE ARCHITECTS OF ARYAMEHR

- What was the introduction that was offered to you regarding the establishment of Aryamehr once you were offered the project? How was the project described to you? What agendas were emphasized to architects?

- Did they have any particular design in mind? Was there any university in the West that they were trying to imitate?
- Hossein Nasr, who was an intellectual architect of Aryamehr, believed science and technology must be adapted to Iranian and Islamic culture. Was there any thought similar to this for the designing of the building?
- Did you also design any other university in Iran at the time? Was there any difference between the way this project was formulated and executed and those other ones?
- The government was concerned with political activism on the campuses. Was this discussed with you? Did the government attempt to translate these concerns into the design of the buildings and the campus?

QUESTIONS TO BE ASKED OF FORMER PROFESSORS OF AMUT

- What do you think of pedagogical methods used in AMUT at the time you were a professor there?
- How did you design the syllabi of your courses? How did other professors design their syllabi?
- Some thinkers such as Dr. Nasr believed science and technology must be adapted to Iranian and Islamic culture. What do you think about this theory? Do you think this influenced teaching and research at AMUT? If so, how?
- Did you have female students at the time? What were the gender policies at AMUT at the time? Were these different from other schools'?
- Did you deal with religion in any aspect related to the university?
- Were you politically active? Did you discuss politics in the classroom?
- How did the political environment affect the research and teaching of the professors?
- What were the topics of research at the time? Were they influenced by the Cold War and White Revolution?

LIST OF ABBREVIATIONS

AMUT	Aryamehr University of Technology
BIPOC	Black, Indigenous, or People of Color
BYU	Brigham Young University
DEI	diversity, equity, and inclusion
Ganjīnah	Sharif Historical Document Center
HOHP	Harvard Oral History Project
ISA	Iranian Student Association
IUT	Isfahan University of Technology
MIT	Massachusetts Institute of Technology
NLAI	National Library and Archives of Iran
OIPFG	Organization of Iranian People's Fedai Guerrillas
PLI	Parliamentary Library of Iran
PMOI	People's Mojahedin Organization of Iran
SACC	Social Action Coordinating Committee
SAVAK	Sāzmān-i 'iṭilā'āt va 'amnīyat-i Kishvar [Bureau for Intelligence and Security of the State]
STEM	science, technology, engineering, and mathematics
USAC	Utah State Agricultural College
UPenn	University of Pennsylvania

GLOSSARY

anjuman 'islāmī dānishjūyān	Islamic Student Association
asāsnāmah	charter
chirīkī	guerrilla militant
fa'ālīyat 'alanī va qānūnī	legal and public activity
'iftikhār	pride
'inqilābī	revolutionary
'itihād, mubārizah, pirūzī	Unity, Struggle, Victory
jav-i khafaqān-i siyāsī	politically suffocating and repressive environment
khalq	oppressed people
kumitah ṣinfī 'ustādān dānishgāh	The faculty union committee of the university
markaz fa'āliyat-hāyi farhangī va fawq barnāmah	The Center for Cultural and Extracurricular Activities
markaz ta'limāt 'umūmī va zabānhāyi Khārijī	Center for Humanities and Foreign Languages
mashy-i chirīkī	manners, social ideals, and worldview of Marxist guerrilla fighters
nāyib tawliyat	Viceroy trustee
nukhbah	talented elite
payām dānishjū	A Message to Students

pishāhang	vanguard
rastākhīz	resurgence
rifāh	welfare
rūz-i dānishjū	Student Day
sāziman tūda'ī	mass organizations
shūrā-yi dānishkadeh	Faculty Council
ṣinfī	guild, union
tawlīyat	trusteeship
vizārat farhang	Ministry of Culture
vizārat 'ulūm va amuzish 'ālī	Ministry of Science and Higher Education

NOTES

INTRODUCTION

1. To read the full report see: https://nfap.com/wp-content/uploads/2022/03/International-Students-in-the-US-and-Canada.DAY-OF-RELEASE.March-2022.pdf

2. Globally, there are several notable examples of engineers who are socially and politically engaged, such as the case of engineers in China (Andreas 2009) and the former Soviet Union (Graham 1993).

3. https://ganjineh.sharif.ir/resources/collection/64/

4 For a selection of the documents accessible at this center see: http://ganjineh .Sharif.ir/home/

5. For a detailed discussion of the student activist numbers and their impact on campus culture, see chapters 4 and 5, respectively.

CHAPTER 1

1. Opinions vary regarding the effectiveness of Iranian state policies and planning for these industrialization projects. Critics of the Shah's regime often contend that these initiatives contributed to the formation of a *ṣanāyi' muntāzh*, or an assembly industry. This was reminiscent of Marxist critics during the Pahlavi era, who argued that the regime served bourgeois interests. "Bourgeois comprador" is a Marxist term used to describe people within the capitalist class in colonial or semi-colonial countries who cooperate with foreign powers, sacrificing their own nation's interests to preserve their economic advantages and power. Conversely, proponents aligned with the Pahlavi regime, including Finance Minister Mohammad Yeganeh, asserted that industrialization was geopolitically and strategically crucial for Iran and had to

start with basic materials such as steel before moving to the stage of manufacturing advanced materials (Yeganeh, interview with HOHP).

2. According to Yeganeh, who was an active participant in the negotiations, the Soviets were not the first choice of partners by the Iranians in this venture. They would have preferred a collaboration with the Germans or Americans, but the Americans not only withheld their support for the project but actively dissuaded German involvement as well. Faced with these circumstances, the Iranian state deftly maneuvered to depict the suggestion for collaboration with the Soviets as originating from the Soviets themselves, while striving to keep the negotiations as discreet as possible. This secrecy stemmed from concerns that the United States might attempt to impede the endeavor (Yeganeh, interview with HOHP).

3. The needs for structural, economic, social, and land reforms can be traced back to the Mosaddegh era. At that time, Mosaddegh had introduced a set of land reforms intended to address the feudalistic aspects of Iran's social and political system (Hassanpour 2022). A decade later, the prime minister of Iran at the time, Ali Amini (in office 1961–1962), contended that resuming Mosaddegh era reforms was essential for tackling the economic challenges prevalent in the nation. Nevertheless, differing perspectives emerged regarding the implementation of these reforms. The Shah ultimately dismissed Ali Amini from his position and assumed a leading role in spearheading what came to be officially celebrated as "the White Revolution" (Amini, interview with HOHP).

4. These tensions will be discussed in detail in chapter 4.

5. After nationalization, the name of the American College was changed to Alborz College.

6. The following articles within the AMUT charter provide valuable insights into the administrative and legal attributes of the university, as well as its relationship with the Shah:

Article 1: Under the tutelage of His Royal Highness Aryamehr, the Aryamehr University of Technology is established in Tehran for an unlimited period.

Article 4: The administration of all university affairs will be under the authority of the Viceroy Trustee (*niyābat tawlīyat 'uẓmá*), who is appointed to this position by the Royal order.

Article 5 (Section 2): The Board of Trustees, appointed by the order of His Royal Highness Aryamehr, is responsible for supervising the university.

Article 10: The Board of Trustees is composed of the Minister of Court, the Minister of Culture, Art, and Education, the Minister of Economy, the Viceroy Trustee, and twenty-five other individuals appointed by the decree of His Highness Aryamehr. After the completion of the first five years, at the end of each year, five members are selected by lottery and may be replaced by new members or retain their positions, as per the decree of Aryamehr.

Article 14: The deans of the faculties will be appointed from among the professors of the respective faculties who have a minimum of ten years of university teaching experience. The appointment will be proposed by the Viceroy Trustee and approved by the Board of Trustees. The term of office for deans is three years, and they may be re-elected.

Article 16: In the event that the viceroy deems it necessary to change the dean of a faculty or a professor for any reason, a report stating the reasons will be submitted to the Board of Trustees. If approved by the Board of Trustees, a new dean or professor will be selected in accordance with the provisions of this charter.

Article 22: The curriculum of each faculty will be established based on the program of a leading similar faculty worldwide. After approval by the university council, it will be communicated to the central council of universities for implementation (*asāsnāmah dānishgāh ṣan'atī āryāmihr*, Ganjīnah Historical Document Center).

7. Exceptions to such economic dependency did exist, such as National University and the Technical College of the National Iranian Oil Company. Pahlavi University, later known as Shiraz University, operated with some autonomy under the nominal control of Pahlavi (Zarghamee 2011). AMUT followed the pattern of Pahlavi University, which, at that time, enjoyed better financial resources due to its proximity to the royal court, as well as fitting within the emerging educational leadership role that the Shah envisioned for himself.

8. Famously, Firooz Partovi, who returned to Iran after graduating from MIT, was arrested by SAVAK. The reason for this was that Partovi returned along with Ali Shariati (1933–1977), famous social critic and former political prisoner, which made SAVAK suspicious of his intentions. However, after more than two months in prison, Partovi was released partly due to the support he received from his former professors at MIT, who wrote letters on his behalf. Upon his release, he joined Polytechnic University before becoming the first professor at AMUT. Partovi served as the first head of the Physics Department and, after a few years, was replaced by Mohammad Hassan Partovi. Firooz Partovi is credited as the first department head whose practices led to the position of department head becoming rotational. Partovi also played an essential role in the early equipping of the laboratories at AMUT (interview with Partovi, 2024).

9. As we will discuss, this earlier tendency to partner with European universities soon shifted toward a greater inclination for US institutions of higher education, once the tenureship of Mojtahedi ended.

10. Firooz Partovi also assisted in recruiting several professors, especially in the field of physics, from the United States (interview with Partovi, 2024).

11. In the winter of 1968, the Ministry of Science and Higher Education (*vizārat 'ulūm va 'āmūzish 'ālī*) was established. Prior to this, universities were under the authority of the Ministry of Culture (*vizārat farhang*). The Ministry of Culture was then divided into the Ministry of Culture and Arts (*vizārat farhang va hunar*) and the Ministry of Science and Higher Education. Following the establishment of the Ministry of Science and Higher Education, the presidents of universities other than royal ones, such as AMUT, came under the supervision of the ministry.

12. Firooz Partovi claims that Mojtahedi's resistance to making such a change was also a reason behind his dismissal (interview with Partovi, 2024).

13. The Welfare Committee was the first Iranian faculty union and played important activist roles in later years, which will be discussed in chapter 4.

14. The name of this center was *Markaz Ta'limāt 'Umūmī va Zabānhāyi Khārijī*, which literally translates to "Center for General Education and Foreign Languages." However, since this literal translation does not fully convey the center's focus, we have opted to use "Center for Humanities" throughout this book when referring to it.

15. These two developments are the subjects of chapters 2 and 3 in the book, respectively.

16. Zarghamee claims that students who were coming from Isfahan were "extremely religious," and they played important roles in the revolution. Describing them as "terrorists," he notes that most of them came from families with a highly political orientation.

17. The success of this symposium was also a result of the efforts by the secretary of the organizing committee, Dr. Firooz Partovi, and Farideh Dowlatshahi. This collaboration was one of the foundations on which AMUT and MIT worked together. A number of students and professors at MIT helped with the preparation of the proceedings of this symposium, which was published in 1973: Feld, Michael S., Ali Javan, and Norman A. Kurnit. "Fundamental and Applied Laser Physics; Proceedings of the Esfahan Symposium, August 29 to September 5, 1971." New York: Wiley, 1973.

18. Introduced by Mahmoud Khayami, a cofounder and owner of the Iran National company and factory at the time.

CHAPTER 2

1. The educational assistance programs initiated during the Truman administration in the 1950s continued to shape educational initiatives until the late 1970s (Shannon 2017). By the 1970s, these initiatives had expanded in scope and pursued more sophisticated objectives. The influx of Iranian students to the US was bolstered by the growth of international education, as well as bilateral security agreements, regional defense pacts, and various forms of cooperation. The focus of these educational relations predominantly revolved around military training, technical education, and cultural exchange (Shannon 2017, 95).

2. An agreement for a program of collaboration between Massachusetts Institute of Technology and Aryamehr University of Technology; see MIT Archives: Box 49 [Barcode: 39080032210368], ASC, C [Range: 30, Bay: 03, Shelf: 03], AC—0397, Massachusetts Institute of Technology, Office of the President and Chancellor, records of Chancellor Paul E. Gray, Series 2.

3. An agreement for a program of collaboration between Massachusetts Institute of Technology and Aryamehr University of Technology, see MIT Archives: Box 49 [Barcode: 39080032210368], ASC, C [Range: 30, Bay: 03, Shelf: 03], AC—0397, Massachusetts Institute of Technology, Office of the President and Chancellor, records of Chancellor Paul E. Gray, Series 2.

4. An agreement for a program of collaboration between Massachusetts Institute of Technology and Aryamehr University of Technology, see MIT Archives: Box 49 [Barcode: 39080032210368], ASC, C [Range: 30, Bay: 03, Shelf: 03], AC—0397, Massachusetts Institute of Technology, Office of the President and Chancellor, records of Chancellor Paul E. Gray, Series 2.

5. On July 17, 1974, in a letter from Chancellor Nasr to MIT President Wiesner, Nasr conveyed that the agreement between the two universities had been signed and returned. In the same letter, Nasr also notes that Wiesner has studied the "shortcomings of the totalitarian Communist regime as far as academic life and scientific activity." Nasr asked Wiesner to send his writings on the issue to him. Given all the ways that Iran's regime at the time was mindful of not building educational relations with the Soviet Union (Ganjavi 2023), it is not surprising that Nasr requested this anti-Communist report in the same letter in which he informed Wiesner about the signing of the agreement between the universities of Iran and America (MIT Archives: Box 49 [Barcode: 39080032210368], ASC, C [Range: 30, Bay: 03, Shelf: 03], AC—0397, Massachusetts Institute of Technology, Office of the President and Chancellor, records of Chancellor Paul E. Gray, Series 2).

6. MIT Archives: Box 49 [Barcode: 39080032210368], ASC, C [Range: 30, Bay: 03, Shelf: 03], AC—0397, Massachusetts Institute of Technology, Office of the President and Chancellor, records of Chancellor Paul E. Gray, Series 2.

7. MIT Archives: Box 7 [Barcode: 39080008701705], HD [HD], AC—0387, Massachusetts Institute of Technology, Office of the President and Chancellor, records of Barbara Scott Nelson, assistant to the president and chancellor.

8. MIT Archives: Box 49 [Barcode: 39080032210368], ASC, C [Range: 30, Bay: 03, Shelf 03], AC—0397, Massachusetts Institute of Technology, Office of the President and Chancellor, records of Chancellor Paul E. Gray, Series 2.

9. MIT Archives: Box 49 [Barcode: 39080032210368], ASC, C [Range: 30, Bay: 03, Shelf: 03], AC—0397, Massachusetts Institute of Technology, Office of the President and Chancellor, records of Chancellor Paul E. Gray, Series 2.

10. From the MIT point of view, this collaboration could take a variety of forms ranging from merely advising to the extent of MIT taking major responsibility of creating and operating a new laboratory and new oceanography program for AMUT.

11. MIT Archives: Box 49 [Barcode: 39080032210368], ASC, C [Range: 30, Bay: 03, Shelf: 03], AC—0397, Massachusetts Institute of Technology, Office of the President and Chancellor, records of Chancellor Paul E. Gray, Series 2.

12. MIT Archives: Box 49 [Barcode: 39080032210368], ASC, C [Range: 30, Bay: 03, Shelf: 03] AC—0397, Massachusetts Institute of Technology, Office of the President and Chancellor, records of Chancellor Paul E. Gray, Series 2.

13. To read more about the history of MIT during the Cold War see: Leslie 1993; Wisnioski 2012; Renehan 2007; Nolan 2018.

14. Weizenbaum was a German-American computer science professor at MIT. He is known as one of the founders of artificial intelligence, as well as one of its earliest critics.

In 1966, he created a program named ELIZA, one of the first chatbots, and the basis for his extensive philosophical statements and criticisms of intelligent technology.

15. MIT Archives: Box 6 [Barcode: 39080008691666], HD [HD], MC—0024, Gordon Stanley Brown papers, Series 2.

16. MIT Archives: Box 6 [Barcode: 39080008691666], HD [HD], MC—0024, Gordon Stanley Brown papers, Series 2.

17. MIT Archives: Box 6 [Barcode: 39080008691666], HD [HD], MC—0024, Gordon Stanley Brown papers, Series 2.

18. By 1976, student activists who had gained experience through their involvement in the Iran case began to express criticism toward another international engagement of MIT, this time involving Taiwan.

19. To review the four years of activism by the Class of 1978, you can refer to *The Tech* newspaper's publication on June 5, 1978.

CHAPTER 3

1. Interview with Hossein Nasr, 2019.

2. Shaikh Fazlolah Nuri (1843–1909), famous Shi'a scholar, was an early supporter of the constitutional movement, who later criticized it for not following the Islamic path he was advocating. He was executed following the abdication of Mohammad Ali Shah and the conquering of Tehran by the constitutionalists. Vanessa Martin, "NURI, FAŻL-ALLĀH," *Encyclopædia Iranica*, online edition, available at http://www .iranicaonline.org/articles/nuri-fazl-allah (accessed on 19 June 2014).

3. MIT Archives: Box 90 [Barcode: 39080020389232], HD [HD], MC—0420, Jerome B. Wiesner papers.

4. MIT Archives: Box 90 [Barcode: 39080020389232], HD [HD], MC—0420, Jerome B. Wiesner papers.

5. MIT Archives: Box 90 [Barcode: 39080020389232], HD [HD], MC—0420, Jerome B. Wiesner papers.

6. MIT Archives: Box 90 [Barcode: 39080020389232], HD [HD], MC—0420, Jerome B. Wiesner papers.

7. MIT Archives: Box 90 [Barcode: 39080020389232], HD [HD], MC—0420, Jerome B. Wiesner papers.

8. MIT Archives: Box 90 [Barcode: 39080020389232], HD [HD], MC—0420, Jerome B. Wiesner papers.

9. MIT Archives: Box 90 [Barcode: 39080020389232], HD [HD], MC—0420, Jerome B. Wiesner papers.

10. To view images of the building's construction, visit: https://www.iut.ac.ir/fa /about-us.

11. MIT Archives: Box 90 [Barcode: 39080020389232], HD [HD], MC—0420, Jerome B. Wiesner papers.

12. MIT Archives: Box 90 [Barcode: 39080020389232], HD [HD], MC—0420, Jerome B. Wiesner papers.

13. MIT Archives: Box 90 [Barcode: 39080020389232], HD [HD], MC—0420, Jerome B. Wiesner papers.

14. MIT Archives: Box 90 [Barcode: 39080020389232], HD [HD], MC—0420, Jerome B. Wiesner papers.

15. For studies on the cultural policy of late Pahlavi era see: chapter 6 in Matin-Asgari 2018; and chapter 4 in Mirsepassi 2019.

16. Gholam Ali Haddad Adel (b. 1945), a conservative Muslim thinker and former affiliate with AMUT, has held several high positions in the government since the revolution of 1979, including chairing the Parliament.

17. William Chittick (b. 1943), is an American professor of Islamic sciences and mysticism.

18. A famous Islamic architectural feature located in the Jameh Mosque of Isfahan, Iran. Constructed in the eleventh century under the Seljuk dynasty, the Taj Al Molk Dome is celebrated for its exceptional structural clarity and intricate decorative craftsmanship.

19. For more information on the history of this Center, see: http://cis.Sharif.ir/introduction/

20. As Nasr explains, "You know that in some Iranian universities, students would go to class for two months each term rather than four months and still they would receive their degrees. Students would not come to class for weeks, and the teacher would claim to make up what they had missed in two days. I had seen that at Tehran University many times, and so it was a great challenge to me to be very strict in the matter of classes at Aryamehr University" (Nasr 2010, 122).

21. These events will be discussed in detail in the next chapter.

CHAPTER 4

1. National Front was an umbrella political network led by Dr. Mohammad Mosad-degh and was organized around the agenda of nationalization of the Iranian oil industry, which had previously been owned by the British-controlled Anglo-Iranian Oil Company (AIOC).

2. For studies on Behrangi see: Khanlarzadeh 2021; Fereshteh 1993.

3. Literally meaning sir, or a man.

4. For an instance of SAVAK's assault at the AMUT dormitory, see Farahmand Rad 2018, 50. In a poignant anecdote, Farahmand Rad recounts an incident in which SAVAK agents raided the AMUT dormitory. The students in the room went to great lengths to hide the political pamphlets belonging to one of their fellow students, believing that they were protecting a comrade. However, in a cruel twist of fate, it was revealed years later that the student they were protecting was a SAVAK agent (Farahmand Rad 2018, 55).

5. Farahmand Rad also describes how, in several instances, it was only years later that he became aware of the political and organizational associations some students had during their time at AMUT (Farahmand Rad 2018, 54).

6. On the occasion of the birth date of the twelfth imam, Morteza Motahhari lectured at AMUT on November 6, 1968, invited by the Islamic association of students.

7. In later years, the organizational accounts of these two groups have been criticized for their possible appropriation of all student activism in favor of their own organization. For instance, see: Itiḥādiyah kumūnīst-hā-yi irān 1975.

8. Vahabzadeh, in his Encyclopedia Iranica article on OIPFG (Vahabzadeh 2015), stated that Masoud Ahmadzadeh (1945–1972), a founding member of OIPFG and a major theorist of the armed movement in Iran, was a mathematics student at AMUT. However, this information is incorrect as Ahmadzadeh was not a student at AMUT but rather an alumnus of STEM education who had studied at the University of Tehran.

9. OIPFG, mubārizah, 'itiḥād, pirūzī, payām bi rufaqāy-i dānishijū bi munāsibat 16 āzar [Struggle, Unity, and Victory: A Message to the Student Comrades on the Occasion of the 16 Azar], December 10, 1979].

10. For more detailed studies see: Vahabzadeh 2010; Rahnema 2021.

11. *Comprador capitalism* is a term used primarily in Marxist economic theory and political critiques to describe a form of capitalism characterized by a dependency relationship between the economic structures of a nation and foreign capital interests. In this system, local capitalists (known as compradors) act as agents or intermediaries for foreign businesses, facilitating the extraction of wealth and resources from the local economy to benefit foreign entities.

12. Although there are other crucial texts from the period of armed struggle, one stands out among the rest: Jazani/Farahani's "What a Revolutionary Should Know." However, this text has not been included in our discussion since its history of publication and circulation suggests that it may not have wielded the same impact on the student body and campus during the time we investigated. Moreover, none of our interviewees mentioned reading this text at the time.

13. For an extended description of the social and economic backgrounds of students at AMUT, see chapter 1.

14. The uncritical approach that the two groups had toward each other has been criticized by Vahabzadeh, arguing that "the Unity of action" resulted in the collapse of the distinction between OIPFG and PMOI militants. Vahabzadeh states, "In the context of an increasingly polarized society, this so-called unity of action through armed struggle gradually clouded the differences between the Left and the religious groups (a difference that must be maintained at all times) and led to the Left's forgetting its own secular foundations" (Vahabzadeh 2011, 92).

15. For instance, Mahdi Amir Shah Karami, member of Muvaḥidīn (موحدین), was AMUT student who tragically lost his life during a military operation. Alireza Shokohi from Sitārah Surkh (ستاره سرخ) and Rāh Kārgar (راه کارگر), and Mohammad Javad Qaedi of 'itiḥād

mubārizah dar rāh ārmān tabaqah kārgar- sahand (اتحاد مبارزه در راه آرمان طبقه کارگر -سهند) also notably continued their activism. To read more, see: https://janbakhteghanerahekargar .wordpress.com/2010/07/14/789/.

Others, like Toraj Heydari Baygvand, formerly associated with OIPFG, became an influential member of a group that broke from OIPFG in 1977 and formed a new group called Munsha'ib (منشعب). According to Vahabzadeh, "In 1977 a major branch consisting of a quarter of the OIPFG, known as the Monsha'ebin, split away from the Fedayeen. They joined the Tudeh Party in March 1979" (Vahabzadeh 2011, 87). For a study of the Munsha'ib group see: Taqizadeh 2021.

16. Five years after its establishment and on the occasion of the first graduates of the university, Shah paid a visit and made some remarks in which he touched upon student unrest as well. Shah noted, "It is normal for youth to make some noise. If they don't it would be unnatural. Some ages have certain characteristics, and if a youth is not full of energy, there should be a health problem with them. . . . There should be noises, but what I hear after the educational revolution is about bus fare, or things like this. It is so nominal that I can't remember. Please someone remind me of such events." Next, the Shah argued that some people have taught that educational revolution means that they don't need to study, or to be examined after a semester, noting, "those who don't study are parasites of a society . . . these are not acceptable to any civilization" (Kārnāmah dah sālah fa'āliyat dānishgāh șan'atī 1977, 161).

17. Paykār, issue 2, 1971.

18. Later, in 1974, the government further ratified a law that would establish a certain quota for the army among the admitted students at the campus. On October 21, 1974, students protested the admission of military students at AMUT. University Guards attacked them. According to Payam dānishjū, the student bulletin of OIPFG, this newly introduced admission policy created a division among students in the first year, as the other students started to position themselves against the students admitted under the military quota. However, from the second year, OIPFG argued that instead of reacting to such students, progressive student activists should attract them, so that it would not be the military that has penetrated the university, but the university that has found a way to penetrate the military (Payam dānishjū, issue 2, 22).

19. Following this terror, SAVAK arrested Mohammad Masoumkhani, a student at AMUT, who died in custody. In a statement issued by OIPFG, the organization denied any involvement by this student in the terror of Sarhang Norouzi (Nabard Khalq, issue 2, 1975; also see: "Darbārah zindigī rafīq shahīd muhammad ma'șūmkhānī" n.d.)

20. One example of this was in response to the execution of Ali Bakeri, former assistant at the chemistry laboratory, and active member of OIPFG of Iran, on April 19, 1972 (Rūzigār sharīf 2016, 116).

21. An example is the demonstrations against Richard Nixon's trip to Iran on May 29, 1972 (Rūzigār sharīf 2016, 120). According to Farahmand Rad, this demonstration was the time students used violence—when they attacked the minibus of the guards with sticks and stones and broke its windows. On the midnight of 16th of Khurdad,

University Guards attacked the AMUT dormitory located in Zanjan Street and started inspecting the student rooms. Farahmand Rad and his roommate were quite frightened, for there was a bag full of (banned) Persian books published by Progress hidden in their room. He recounts that the most dangerous of all these items was Maxim Gorky's *Mother*. Farahmand Rad lists Alex Haley's 1976 novel *Roots: The Saga of an American Family* as one of the books in the bag. However, it seems the dates do not correspond, as this novel was not published until a few years later (Farahmand Rad 2018).

22. October 16, 1972.

23. On October 21, 1974, students protested the admission of military students at university. The protests continued the next day. University Guards attacked them.

24. As noted by Taghi Shahram in the Tapes of Conversation between PMOI and OIPFG. Accessible at: https://www.peykarandeesh.org/PeykarArchive/Mojahedin-ML /mojahed_fadaii.html.

25. The Tapes of Conversation between PMOI and OIPFG. Accessible at: https:// www.peykarandeesh.org/PeykarArchive/Mojahedin-ML/mojahed_fadaii.html

26. OIPFG's legendary operational leader, Hamid Ashraf (1946–1976), was killed in an attack by the police. A few AMUT students, who had become guerrilla fighters, were also among those killed during this attack at Ashraf's safe house, including Ali Akbar Vaziri, a member of the AMUT mountain climbing group and member of OIPFG. Additionally, Tahereh Khorram, a student of mechanics at AMUT, was also killed in this police raid. For more information, see: Ebrahmizadeh 2021.

27. The SAVAK agent suggested that more police should closely observe Eisenhower Street. SAVAK documents, Report number 1102/15401.

28. These poetry nights, a series of poetry and essay readings held between October 10–19, 1977, were a collaboration between Iran's Writers Association and the Irano-German Goethe-Institut in Iran. They soon became highly politicized, with thousands of attendees expressing antiregime sentiments. This event marked one of the beginnings of mass demonstrations against the Shah regime (Karimi-Hakkak 1985).

29. This was just a few days after November 15, 1977, during which students at AMUT, along with some faculty, did not attend classes in protest to the students who had been detained. Interestingly, the next day, the news of this was passed on to students at University of Tehran, while SAVAK agents were closely monitoring how this would evolve (SAVAK documents, Report number 1102/15316).

30. About ten days before this event, 300 students at University of Tehran were beaten by riot police in a conference hall. Student cards of 150 students were confiscated, to which the students reacted by boycotting classes.

31. Another letter from the US embassy in Tehran to the secretary of state in Washington (dated July 1978) provides further details on the trial held for student activists. The press reported on July 5 and 6 that courts fined two AMUT students 30,000 rilas each (about $425) on charges of "disturbing public order while demonstrating on streets." The court further freed on bail forty students accused of "unruly demonstrations" in Tehran.

Also a letter from Pahlavi University faculty to the prime minister has been the "first public outside support" AMUT picked up in its campaign to keep the Tehran campus open. The argument in the letter was that this campus was producing 7,000 engineers per year, in a country that needs 50,000 engineers over the coming years (Letter from American Embassy in Tehran to Secretary of State, Subject: Political Activities, July 1978).

32. The Isfahan campus commenced operations this year, 1977. Even so, very soon this new campus also became the site of protests. On December 4, 1977, students at the Isfahan campus demonstrated. Ten students were detained (*Ruzigar Sharif* 2016, 170).

33. Zarghamee, in his interview with Harvard Oral History, notes that in his last audience with the Shah, he mentioned to the Shah about the increase in Islamic opposition among AMUT students, arguing that from what he could see, the Shah should be more cautious about Islamic opposition rather than Marxist opposition, which he had found to be relatively much less powerful at the time. He recounts that the Shah rejected any threat from the so-called religious sentiments, pointing to how the vast development of television coverage in Iran will soon help with altering such sentiments (Zarghamee, interview with HOHP).

34. The reduction of salary for those professors who were protesting the moving of the AMUT to Isfahan has also been discussed in a conversation between Mehdi Bazargan, member of The Freedom Movement of Iran, and John D. Stempel from the American embassy in Iran on May 30, 1978. In this conversation, Bazargan further says that he thought the army was intending to use the campus space to establish its own engineering school.

35. In the term "clerical strike," Khansari refers to a strike conducted by staff members of a department, institution, or governmental office, rather than by workers in a company.

36. Iran Writers Association expressed its support for the AMUT strike in a declaration on May 8, 1978.

37. The *Rastākhīz* newspaper, the organ of the Rastākhīz Party, wrote an editorial against the faculty members on December 7, 1978/16 Azar. Members of Atomic Energy Organization of Iran, the Polytechnique faculty, Ferdowsi University, Teacher Training University, Jondi Shapur University, Pahlavi University, Bu-Ali Sina University, and many others all signed petitions in support of AMUT's students and professors.

38. It should be noted that before this year, the entrance exam was administered by the Ministry of Higher education and not the university.

39. Sayf Kurdi also narrates that his name was on a list of thirty-eight people who SAVAK intended to arrest, but this failed with the revolution. He notes the reason for that was that he was the person in charge of enrolling new students.

40. For a report on these events, see: *Washington Post*, "The Shah as Tyrant: A Look at the Record," March 20, 1980.

41. Another SAVAK document dated December 23, 1978, from the Mazandaran branch of SAVAK to the head of the third branch of SAVAK, notes that the police

in Behshahr city stopped a car in which a student of mechanics at AMUT, named Morteza Muhavilati, along with another student from Madrisa Aali Bazargani, were carrying images of Khomeini, three copies of a pamphlet titled "Artish Muzdur Iran," some books including Marx and Engels's *The Communist Manifesto*, as well as other books on historical materialism and dialectical materialism.

42. Between 1971 and 1979, 237 OIPFG members lost their lives. About 55 percent of these casualties occurred in street battles, while nearly 20 percent were executed by firing squad. More than 80 percent of all Fedayee deaths, including executions, occurred in Tehran. Women accounted for more than 15 percent of the Fedayee casualties. As the new theocratic state consolidated its power, the OIPFG emerged as Iran's largest leftist group. However, it soon lost popularity due to multiple splits within the group, severe repression by the Islamic state, and the forced migration of thousands of leftist activists. Consequently, the OIPFG has had little effective presence in Iranian society since the mid-1980s (Vahabzadeh 2011).

CHAPTER 5

1. For a report see: *Radio Zamaneh*, April 13, 2024. Accessible at: https://www.radio zamaneh.com/812656.

2. Polytechnic University has been and remains one of the most prominent universities in the field of STEM education, alongside the University of Tehran and AMUT.

3. The number of female attendees each year never exceeded 10 percent of the total. For detailed statistics on female students at AMUT, see chapter 1.

4. *Quartz* "The West Is Way behind Iran and Saudi Arabia When It Comes to Women in Science," March 8, 2018. Also see: Zahidi 2018.

5. Khanlarzadeh explains in her engagement with Jack Halberstam's theories of female masculinity (Halberstam 1998), the term references, "women who adopt behavior conventionally associated with men and their lifestyles in a given socio-historical context" (Khanlarzadeh 2022, 2).

6. The student activism during the chancellorship of Nasr has been discussed in detail in the previous chapter.

CHAPTER 6

1. While similar in name and surface characteristics, we note significant ethical and philosophical differences between nativist scientific thought in Iran and the diverse and heterogeneous scientific traditions of Native Americans in the Americas.

2. We emphasize here that students' rejection of Islamic thought within the discipline of engineering, or at AMUT more broadly, is not an indicator of any individuals' religious background or affiliation. As described in chapters 4 and 5, many among AMUT students had Islamic backgrounds.

3. He currently lives in Tehran, where he is a widely recognized and respected public intellectual. His writings often focus on subjects directly relevant to the heart of our inquiry here—the dynamic between technology, modernization, politics and

inequality. Among several other works, he has published monographs titled *Introduction to Critical Thinking* (Akhtaran 2014), *Authoritarianism in Iran* (Akhtaran 2008), and *A Treatise on the Social Psychology of Iranian People* (Akhtaran 2009).

Ghazimoradi, Ḥassan. Darāmadī bar tafakkur-i intiqādī. Tihrān: Intishārāt-i Akhtarān, 1393 [2014].

Ghazimoradi, Ḥassan, Darāmadī bar tafakkur-i intiqādī (Tihrān: Intishārāt-i Akhtarān, 1393 [2014]).

Ghazimoradi, Ḥassan. Dar pīrāmūn-i khawdmadārī-yi īrāniyān: Risālah-ī dar rāvānshināsī-yi ijtimāʿī-yi mardum-i īrān. Tihrān: Intishārāt-i Akhtarān, 1387 [2008].

4. Black studies scholar Sadiya Hartman developed critical fabulation as an interpretive method that forwards artistic creation as a form of historical inquiry (Hartman 2008). In her recent book, *Wayward Lives, Beautiful Experiments: Intimate Histories of Social Upheaval* (2019), Hartman uses critical fabulation by mobilizing the archive to construct artistic renditions of the intimate lives of Black women in the early twentieth century. In this way, critical fabulation resonates with interpretive traditions in social science research that combine art and science, such as portraiture (Lawrence-Lightfoot 2005), narrative inquiry (Clandinin 2016), and speculative fiction methods (Toliver 2020), as well as new directions in the learning sciences which explore the aesthetic, emotive, and poetic dimensions of human learning (Vossoughi 2021).

5. We extend Hartman's use of archive to include oral histories as an additional source of inspiration for the development of the vignettes. We interviewed Hassan on three separate occasions, amounting to nearly seven hours of interview data. The interviews were conducted in Persian, then translated into English and carefully analyzed for themes relevant to our research question. As an example, in vignette 3, we describe a scene in which Hassan's interlocutor is his professor, legendary poet Esmail Khoei. The vignette describes a conversation between Hassan and Khoei as they weave through the notoriously congested traffic of Tehran. Here, the car trip is fictional, but the conversation is based directly on the interview transcript in which Hassan describes several off-the-record conversations with Khoei, who actually was his professor.

6. We should note that we are aware that the term "Iranianizing" is vague and has been interpreted differently by various groups. As demonstrated in various chapters of the book, when the Shah aimed to "Iranianize" the university, this encompassed both adapting the university to address the particular economic and industrial demands of Iranian society at that time and also entailed a willingness to embrace initiatives that aimed to integrate the study of Iran's culture, including Islam, into the curriculum.

CONCLUSION

1. From Oral Histories. Interview conducted on May 14, 2022.

EPILOGUE

1. Alborz College introduced innovative modes and forms of training and pedagogy, leaving an indelible mark on the country's education landscape. Notably, it educated

political elites of the time, enrolling the sons of royalty, prime ministers, cabinet members, tribal chieftains, and influential figures from across the country. The institution's influence extended beyond academics, shaping the cultural and social landscape with initiatives like scouts, summer camps, and various extracurricular activities. The extensive history of the American College of Tehran unfolds across three distinct periods: its establishment until the 1940s, the era spanning from the 1940s to 1979, and the postrevolutionary period. Two monumental figures, Samuel Jordan (1899–1940) and Muhammad Ali Mujtahidi (1944–1979), served as managers and directors, significantly influencing the institution's development and earning praise from former students. To study this institution, we have a wealth of sources at our disposal, including oral histories, archives at the Presbyterian Historical Society, papers from former professors like Walter Alexander Groves, and correspondence with Iran's Ministry of Education stored at Iran's national library.

BIBLIOGRAPHY

Abdi, Abbas. *Junbish-i dānishūj'ī pulītiknīk tihrān (1338–1357)*. Tehran: Nay, 2014.

Abrahamian, Ervand. *Iran between Two Revolutions*. Princeton, NJ: Princeton University Press, 1982.

Abrahamian, Ervand. *The Iranian Mujahedin*. New Haven: Yale University Press, 1989.

Abrahamian, Ervand. "The 1953 Coup in Iran." *Science & Society* 65, no. 2 (2001): 182–215. http://www.jstor.org/stable/40403895.

Afary, Janet. *Sexual Politics in Modern Iran*. Cambridge; Cambridge University Press, 2009.. https://doi.org/10.1017/CBO9780511815249.010.

Ahmadzadeh, Masoud. *Mubārizah musalahānah: Ham 'istiratizhī, ham tāktīk*. n.p., 1970.

Ahmed, Sara. *On Being Included: Racism and Diversity in Institutional Life*. Durham, NC: Duke University Press, 2020.

Aikenhead, Glen S. *Science Education for Everyday Life: Evidence-Based Practice*. New York: Teachers College Press, 2006.

Al-Hassani, Salim T. S. *1001 Inventions: The Enduring Legacy of Muslim Civilization*. 3rd ed. Washington, DC: National Geographic, 2012.

Alam, Assadullah. *Yāddāsht-hāyi 'alam*. Edited by Alinaghi Alikhani. Tehran: Kitabsara, 2011.

Amirkhosravi, Babak. "Muṣāḥibah bā Bābak Amīrkhusrawī" YouTube video. Posted [2015]. https://www.youtube.com/watch?v=Tkr5vv-m11c

Amirkhosravi, Babak. *Zindigīnāmah sīyāsī*. Sweden: Baran, 2020.

Anderson, Carol. *The Second Race and Guns in a Fatally Unequal America*. New York: Bloomsbury Publishing, 2021.

Andreas, Joel. *Rise of the Red Engineers: The Cultural Revolution and the Origins of China's New Class*. Stanford, CA: Stanford University Press, 2009.

Anteby, Michel. *Moral Gray Zones: Side Productions, Identity, and Regulation in an Aeronautic Plant*. Princeton: Princeton University Press, 2008.

Aquilina, Susannah. "Common Ground: Iranian Student Opposition to the Shah on the US/Mexico Border." *Journal of Intercultural Studies* 32, no.4 (2011): 321–334.

Armajani, Yahya. "Alborz College." In *Encyclopaedia Iranica*. 1985. Last accessed January 10, 2020. www.iranicaonline.org/articles/alborz-college.

Aryafar, Kamelia. "How Iranian Immigrants Can Be Role Models for Diversity in STEM." *The Hill*, February 5, 2020. https://thehill.com/changing-america/opinion/481684-how-iranian-immigrants-can-be-role-models-for-diversity-in-stem/#:~:text=That%20culture%20has%20opened%20the,vastly%20underrepresented%20in%20STEM%20fields.

Baghi, Emad al-Din. *Junbish-i dānishjū'ī īrān, az āqāz tā inqilāb 'islāmī*. Tehran: Jami'eh Iranian, 1999.

Bang, M. (2020). "Learning on the Move toward Just, Sustainable, and Culturally Thriving Futures." *Cognition and Instruction* 38, no. 3: 434–444.

Bang, Megan, and Douglas Medin. "Cultural Processes in Science Education: Supporting the Navigation of Multiple Epistemologies." *Science Education (Salem, Mass.)* 94, no. 6 (2010): 1008–1026. https://doi.org/10.1002/sce.20392.

Bang, Megan, Lawrence Curley, Adam Kessel, Ananda Marin, Eli S. Suzukovich, and George Strack. "Muskrat Theories, Tobacco in the Streets, and Living Chicago as Indigenous Land." *Environmental Education Research* 20, no. 1 (2014): 37–55. https://doi.org/10.1080/13504622.2013.865113.

Barajas-López, F., & Bang, M. (2018). "Indigenous Making and Sharing: Claywork in an Indigenous STEAM Program." *Equity & Excellence in Education* 51, no, 1: 7–20.

Barone, Thomas E. "Beyond Theory and Method: A Case of Critical Storytelling." *Theory into Practice* 31, no. 2 (1992): 142–146. https://doi.org/10.1080/00405849209543535.

Barton, Angela Calabrese. *Teaching Science for Social Justice*. New York: Teachers College Press, 2003.

Baser, Bahar, Samim Akgönül, and Ahmet Erdi Öztürk. "'Academics for Peace' in Turkey: A Case of Criminalising Dissent and Critical Thought via Counterterrorism Policy." *Critical Studies on Terrorism* 10, no. 2 (2017): 274–296. https://doi.org/10.1080/17539153.2017.1326559.

Bassett, Ross Knox. *The Technological Indian*. Cambridge, MA: Harvard University Press, 2016.

Behazin, Mahmood Etemadzadeh. *Az har darī*. Tehran: Jami, 1993.

Behrooz, Maziar. *Rebels with a Cause: The Failure of the Left in Iran*. London: I.B. Tauris, 2000.

Blickenstaff, J. (2005). "Women and Science Careers: Leaky Pipeline or Gender Filter?" *Gender and Education* 17, no. 4: 369–386.

Boroujerdi, Mehrzad. *Iranian Intellectuals and the West: The Tormented Triumph of Nativism*. 1st ed. Syracuse, NY: Syracuse University Press, 1996.

Branigin, William. "Shah's Opponents Beaten by Mob in Iran." *The Washington Post*, November 22, 1977.

Brickhouse, N. W., Lowery, P., & Schultz, K. (2000). "What Kind of a Girl Does Science? The Construction of School Science Identities." *Journal of Research in Science Teaching: The Official Journal of the National Association for Research in Science Teaching* 37, no. 5: 441–458.

Butler, Judith. *Gender Trouble: Feminism and the Subversion of Identity*. New York: Routledge, 1990.

Carlone, Heidi B., and Angela Johnson. "Understanding the Science Experiences of Successful Women of Color: Science Identity as an Analytic Lens." *Journal of Research in Science Teaching* 44, no. 8 (2007): 1187–1218. https://doi.org/10.1002/tea.20237.

Choudry, Aziz. *Learning Activism: The Intellectual Life of Contemporary Social Movements*. Toronto: University of Toronto Press, 2015.

Choudry, Aziz, and Salim Vally, eds. *The University and Social Justice: Struggles across the Globe*. London: Pluto Press, 2020.

Clandinin, D. Jean. *Engaging in Narrative Inquiry*. Abingdon, Oxon: Routledge, 2016. https://doi.org/10.4324/9781315429618.

Cole, Michael. *Cultural Psychology: A Once and Future Discipline*. Cambridge, MA: Belknap Press of Harvard University Press, 1996.

Collier-Thomas, Bettye, and Vincent P. Franklin. *Sisters in the Struggle: African American Women in the Civil Rights-Black Power Movement*. New York: New York University Press, 2001.

Curnow, Joe, Amil Davis, and Lila Asher. "Politicization in Process: Developing Political Concepts, Practices, Epistemologies, and Identities Through Activist Engagement." *American Educational Research Journal* 56, no. 3 (2019): 716–752. https://doi.org/10.3102/0002831218804496.

d'Ambrosio, Ubiratan. "Ethnomathematics and Its Place in the History and Pedagogy of Mathematics." *For the Learning of Mathematics* 5, no. 1 (1985): 44–48.

"Darbārah zindigī rafīq shahīd muḥammad ma'ṣūmkhānī." n.d. Accessed April 29, 2024. https://siahkal.com/%D8%AF%D8%B1%D8%A8%D8%A7%D8%B1%D9%87-% D8%B2%D9%86%D8%AF%DA%AF%DB%8C-%D8%B1%D9%81%DB%8C%D9%82 -%D8%B4%D9%87%DB%8C%D8%AF-%D9%85%D8%AD%D9%85%D8%AF-%D9% 85%D8%B9%D8%B5%D9%88%D9%85%D8%AE%D8%A7%D9%86/

Debray, Régis. *Revolution in the Revolution? Armed Struggle and Political Struggle in Latin America.* New York: MR Press, 1967.

Dollimore, Jonathan. "Desire and Difference: Homosexuality, Race, Masculinity." In *Race and the Subject of Masculinities,* edited by Harry Stecopoulos and Michael Uebel, 17–44. New York, USA: Duke University Press, 2020. https://doi.org/10.1515 /9780822397748-003.

Dorraj, Manochehr. "The Political Sociology of Sect and Sectarianism in Iranian Politics: 1960–1979." *Journal of Third World Studies* 23, no. 2 (2006): 95–117.

Downey, Gary Lee, and Juan C. Lucena. "National Identities in Multinational Worlds: Engineers and 'Engineering Cultures.'" *International Journal of Continuing Engineering Education and Lifelong Learning* 15, no. 3–6 (2005): 252–260.

Du Bois, William Edward Burghardt. *The Talented Tenth.* New York, NY: James Pott and Company, 1903.

Dumas, Michael J., and Kihana Miraya Ross. "Be Real Black for Me: Imagining Black-Crit in Education." *Urban Education (Beverly Hills, Calif.)* 51, no. 4 (2016): 415–442. https://doi.org/10.1177/0042085916628611.

Ebrahimzadeh, Mehdi. *Hamrāhān shafīq.* 2021. Accessed April 29, 2024. https://asre -nou.net/php/view.php?objnr=51784.

Elling, Rasmus. "'In a Forest of Humans': The Urban Cartographies of Theory and Action in 1970s Iranian Revolutionary Socialism." In *Global 1979: Geographies and Histories of the Iranian Revolution,* edited by Arang Keshavarzian and Ali Mirsepassi, 141–77. Cambridge: Cambridge University Press, 2021. https://doi.org/10.1017 /9781108979658.010.

Emerson, Robert M., Rachel Fretz, and Linda Shaw. "In the Field: Participating, Observing, and Jotting Notes." *Writing Ethnographic Fieldnotes* 2 (1995): 17–35.

Erickson, Frederick. "Qualitative Methods in Research on Teaching." In *Handbook of Research on Teaching,* edited by Merlin Carl Wittrock, 119–61. New York: Macmillan, 1986.

Esmonde, Indigo, and Angela N. Booker, eds. *Power and Privilege in the Learning Sciences: Critical and Sociocultural Theories of Learning.* New York: Routledge, 2017.

Farahmand Rad, Shiva. *Qatrān dar 'asal.* 2nd ed. Sweden: Self-published, 2018.

Farahmand Rad, Shiva. "Zindigī chirikī dar khāna-hā-yi tīmī." *I Don't Know . . . chi mīdānam,* 2019. Accessed April 29, 2024. https://shivaf.blogspot.com/2019/10/madi -nomreh-20.html

Farahmand Rad, Shiva. "I Don't Know . . . chi mīdānam . . . ." *Zindagī-yi chirkī dar khānah-hā-yi tīmi.* Accessed September 1, 2023. https://shivaf.blogspot.com/2019/.

Fassihi, Farnaz. "'Geniuses' Versus the Guns: A Campus Crackdown Shocks Iran." *New York Times,* October 6, 2022.

Fawcett, Louise. "Revisiting the Iranian Crisis of 1946: How Much More Do We Know?" *Iranian Studies* 47, no. 3 (2014): 379–399. https://doi.org/10.1080/00210862 2014.880630.

Feld, Michael S., Ali Javan, and Norman A. Kurnit. "Fundamental and Applied Laser Physics: Proceedings of the Esfahan Symposium, August 29 to September 5, 1971." New York: Wiley, 1973.

Fereshteh, M. Hossein. "International Rural Education Teachers and Literacy Critics: Samad Behrangi's Life, Thoughts, and Profession." Paper Presented at the Annual Conference of Comparative and International Education Society. Retrieved from https://files.eric.ed.gov/fulltext/ED364542.pdf (accessed January 10, 2020).

Gambetta, Diego, and Steffen Hertog. *Engineers of Jihad: The Curious Connection between Violent Extremism and Education.* Princeton: Princeton University Press, 2018.

Ganjavi. Mahdi. *Education and the Cultural Cold War in the Middle East: The Franklin Book Programs in Iran.* London: I.B. Tauris, 2023

Ganjavi, Mahdi, and Shahrzad Mojab. "A Lost Tale of the Student Movement in Iran." In *Reflections on Knowledge, Learning and Social Movements,* edited by Aziz Choudry and Salim Vally, 55–69. Abingdon: Routledge, 2018. https://doi.org/10 .4324/9781315163826-4.

Garibay, Juan C. "STEM Students' Social Agency and Views on Working for Social Change: Are STEM Disciplines Developing Socially and Civically Responsible Students?" *Journal of Research in Science Teaching* 52, no. 5 (2015): 610–632. https://doi .org/10.1002/tea.21203.

Garlitz, Richard. "U.S. University Advisors and Education Modernization in Iran, 1951–1967." In *Teaching America to the World & the World to America: Education and Foreign Relations since 1870,* edited by Richard Garlitz and Lisa Jarvinen, 195–216. New York: Palgrave Macmillan, 2012.

Ghazimoradi, Hassan. *Istibdād dar īrān.* Tehran: Akhtaran, 2008.

Ghazimoradi, Hassan. *Dar pirāmūn khudmadārī īrāniyān: Risālah-'ī dar ravānshināsī ijtimā'ī mardum īrān.* Tehran: Akhtaran, 2009.

Ghazimoradi, Hasan. *Darāmadī bar tafakur intiqādī.* Tehran: Akhtaran, 2014.

Gheissari, Ali, ed. *The American College of Tehran: A Memorial Album.* Irvine, CA: Jordan Center for Persian Studies and University of California, 2020.

Gillmor, C. Stewart. *Fred Terman at Stanford: Building a Discipline, a University, and Silicon Valley.* Stanford: Stanford University Press, 2004.

Giroux, Henry A. *University in Chains: Confronting the Military-Industrial-Academic Complex*. New York: Routledge, 2015.

Graham, Loren R. *The Ghost of the Executed Engineer: Technology and the Fall of the Soviet Union*. Cambridge, MA: Harvard University Press, 1993.

Gupte, Pranay. "U.S. job seekers looking to Iran." *New York Times*, April 20, 1975.

Gutiérrez, Kris D., and Barbara Rogoff. "Cultural Ways of Learning: Individual Traits or Repertoires of Practice." *Educational Researcher* 32, no. 5 (2003): 19–25. https://doi.org/10.3102/0013189x032005019.

Gutiérrez, Kris D., Patricia Baquedano-López, and Carlos Tejeda. "Rethinking Diversity: Hybridity and Hybrid Language Practices in the Third Space." *Mind, Culture and Activity* 6, no. 4 (1999): 286–303. https://doi.org/10.1080/10749039909524733.

Hakim, Maryam. *Education and Modernization in Iran: Planning and Impact of Educational Policy*. Unpublished PhD diss., State University of New York, 1979.

Halberstam, Jack. *Female Masculinity*. Durham: Duke University Press, 1998. https://doi.org/10.1515/9780822378112.

Harding, Sandra G. *The Postcolonial Science and Technology Studies Reader*. Durham: Duke University Press, 2011.

Hartman, Saidiya. "Venus in Two Acts." *Small Axe: A Journal of Criticism* 12, no. 2 (2008): 1–14. https://doi.org/10.1215/-12-2-1.

Hartman, Saidiya V. *Wayward Lives, Beautiful Experiments: Intimate Histories of Social Upheaval*. 1st ed. New York: W.W. Norton & Company, 2019.

Hassanpour, Amir. *'Ariyanpūr wa jāmi'ahshināsī mārksīstī: Tārīkh, tabaqah ijtimā'ī va diyāliktīk*. Toronto: Irannamag, 2021.

Hassanpour, Amir. *Shūrish-i dahqānān-i Mukrīyān, 1331–1332 sh / 1952–1953 m: Asnād-i kunsūlgarī mukātabāt-i dīplumātīk va guzārish-i rūznāmah'hā*. Toronto: Asemana Books, 2022.

Haqshinas, Torab. *Az fiyziyah tā paykār*. Frankfurt: Andeesheh va Peykar Publications, 2020.

Heffernan, Anne, and Noor Nieftagodien, eds. *Students Must Rise: Youth Struggle in South Africa before and beyond Soweto '76*. Johannesburg: Wits University Press, 2016.

Hess, Diana E., and Paula McAvoy. *The Political Classroom: Evidence and Ethics in Democratic Education*. New York, NY: Routledge, 2015.

Hodgkinson, Dan, and Luke Melchiorre. "Africa's Student Movements: History Sheds Light on Modern Activism." *The Conversation*, February 18, 2019. http://theconversation.com/africas-student-movements-history-sheds-light-on-modern-activism-111003.

Hogendijk, Jan P., and Abdelhamid I. Sabra. *The Enterprise of Science in Islam: New Perspectives*. Cambridge, MA: MIT Press, 2003.

Hojabri, Fereydon. *Bar āb va ātash. zindigī va khāṭirat duktur firaydūn huzhabrī*. Edited by Goel Cohen. Tehran: Kavir, 2016.

hooks, bell. *Teaching Community: A Pedagogy of Hope*. New York: Routledge, 2003.

"Infijār maqar gārd dānishgāh ṣan'atī [Blowing Up the Campus Police Station at San'ati University]." PMOI, 1974.

ʾtiḥādiyah kumūnīst-hā-yi irān. *Junbish dānishjū'ī irān (yik barrisī mukhtasar)*. n.p., 1975.

Jackson, Ava, Lauren Vogelstein, Heather Clark, Lindsay Lindberg, Noami Thompson, and Suraj Uttamchandani. "Learning Together: Reflections at the Intersection of Friendship, Research, and Learning Processes." In *The Interdisciplinarity of the Learning Sciences, 14th International Conference of the Learning Sciences (ICLS) 2020*, edited by Melissa Gresalfi and Ilana Seidel Horn, Vol. 2, 657–60. Nashville, TN: International Society of the Learning Sciences, 2020.

Jahani Asl, Mohammad Nasser. *A Democratic Alternative Education System for Iran: An Historical and Cultural Study*. Unpublished Master's thesis, Simon Fraser University, 2007.

Jazani, Bijan. *Tārīkh sī sālah*. n.p., n.d.

Jurow, A. Susan, and Molly Shea. "Learning in Equity-Oriented Scale-Making Projects." *The Journal of the Learning Sciences* 24, no. 2 (2015): 286–307. https://doi.org/10.1080/10508406.2015.1004677.

Kaiser, David, ed. *Becoming MIT: Moments of Decision*. Cambridge, MA: MIT Press, 2010.

Kamaly, Hossein. "The Cold War and Education in Science and Engineering in Iran, 1953–1979." In *Global 1979: Geographies and Histories of the Iranian Revolution*, edited by Arang Keshavarzian and Ali Mirsepassi, 328–54. The Global Middle East. Cambridge: Cambridge University Press, 2021.

Kang, Hosun, Angela Calabrese Barton, Edna Tan, Sandra Simpkins, Hyang-yon Rhee, and Chandler Turner. "How Do Middle School Girls of Color Develop STEM Identities? Middle School Girls' Participation in Science Activities and Identification with STEM Careers." *Science Education (Salem, Mass.)* 103, no. 2 (2019): 418–39. https://doi.org/10.1002/sce.21492.

Karimi-Hakkak, Ahmad. "Protest and Perish: A History of the Writers' Association of Iran." *Iranian Studies* 18, no. 2–4 (1985): 189–229. http://www.jstor.org/stable/4310495.

Kārnāmah ṣah sālah fa'āliyat dānishgāh ṣan'atī. Tehran: Daftar matbū'āṭī dānishgāh ṣan'atī āryārnihr, 1977.

Kashani-Sabet, Firoozeh. *Heroes to Hostages: America and Iran, 1800–1988*. Cambridge, United Kingdom: Cambridge University Press, 2023.

Katouzian, Homa. "Alborz and Its Teachers." *Iranian Studies* 44, no. 5 (2011): 743–754. https://doi.org/10.1080/00210862.2011.570483.

Kenney, Martin. *Understanding Silicon Valley: The Anatomy of an Entrepreneurial Region*. Stanford, CA: Stanford University Press, 2000.

Khanlarzadeh, Mina. "Theology of Revolution: In Ali Shari'ati and Walter Benjamin's Political Thought." *Religions* 11, no. 10 (2020): 504. https://doi.org/10.3390/rel11100504.

Khanlarzadeh, Mina. "The Stories of Rebellious Children at the Time of the 1979 Revolution." *British Journal of Middle Eastern Studies* 50, no. 2 (2021): 450–464. https://doi.org/10.1080/13530194.2021.1978280.

Khanlarzadeh, Mina. "'More Champion than the Champions': Female Masculinity in *Lālehzari* Music and Filmfarsi." *Popular Music and Society* 46, no. 1 (2022): 1–20. https://doi.org/10.1080/03007766.2022.2117977.

Kifner, John. "Iranian Program Debated and M.I.T." *New York Times*, April 27, 1975. Available at https://www.nytimes.com/1975/04/27/archives/iranian-program-debated-at-mit-training-of-atom-scientists-called.html [Accessed February 7, 2024].

Kline, Ronald R. *Steinmetz: Engineer and Socialist*. Baltimore: Johns Hopkins University Press, 1992.

Kuutti, Kari. "Activity Theory as a Potential Framework for Human-Computer Interaction Research." In *Context and Consciousness: Activity Theory and Human Computer Interaction*, edited by Bonnie A. Nardi, 17–44, Cambridge: MIT Press, 1996.

Lagrange, Hugues. "Diego Gambetta and Steffen Hertog: Engineers of Jihad: The Curious Connection between Violent Extremism and Education." *European Sociological Review* 33, no. 1 (2017): https://doi.org/10.1093/esr/jcw040.

Lave, Jean. "The Practice of Learning." In *Understanding Practice*, edited by Seth Chaiklin and Jean Lave, 3–32. Cambridge University Press, 1993. https://doi.org/10.1017/CBO9780511625510.002.

Lawrence-Lightfoot, Sara. "Reflections on Portraiture: A Dialogue Between Art and Science." *Qualitative Inquiry* 11, no. 1 (2005): 3–15. https://doi.org/10.1177/1077800404270955.

Layton, Edwin T. *The Revolt of the Engineers: Social Responsibility and the American Engineering Profession*. Baltimore: Johns Hopkins University Press, 1986.

Lee, Carol D. "The Centrality of Culture to the Scientific Study of Learning and Development: How an Ecological Framework in Education Research Facilitates Civic Responsibility." *Educational Researcher* 37, no. 5 (2008): 267–279. https://doi.org/10.3102/0013189X08322683.

Lee, Clifford, and Elisabeth Soep. *Code for What?: Computer Science for Storytelling and Social Justice*. Cambridge, MA: The MIT Press, 2023.

Lehrich, Mark Jonathan. "A Matter of Science: The Massachusetts Institute of Technology and the Transformation of American Management Education, 1950–1964." PhD diss., ProQuest Dissertations Publishing, 2016.

Leslie, Stuart W. *The Cold War and American Science: The Military-Industrial-Academic Complex at MIT and Stanford*. New York: Columbia University Press, 1993.

Leslie, Stuart W., and Robert Kargon. "Exporting MIT: Science, Technology, and Nation-Building in India and Iran." *Osiris (Bruges)* 21, no. 1 (2006): 110–130. https://doi.org/10.1086/507138.

Li Zhuying. "Female Masculinity and the Image of Women in the Chinese Cultural Revolution." The Asian Conference on Media and Mass Communication 2014 Official Conference Proceeding (2014), 203–214.

Ma Leanne, Yoshiaki Matsuzawa, Bodong Chen, and Marlene Scardamalia. "Community Knowledge, Collective Responsibility: The Emergence of Rotating Leadership in Three Knowledge Building Communities." In *Transforming Learning, Empowering Learners: The International Conference of the Learning Sciences (ICLS) 2016*, Vol. 1, edited by Chee-Kit Looi, Joseph Polman, Ulrike Cress, and Peter Reimann. Singapore: International Society of the Learning Sciences, 2016.

Marashi, Afshin. *Nationalizing Iran: Culture, Power, and the State, 1870–1940*. Seattle, WA: University of Washington Press, 2008.

Margolis, Jane. *Stuck in the Shallow End: Education, Race, and Computing*. Updated edition. Cambridge, MA: MIT Press, 2017.

Martin, Diana Adela, Eddie Conlon, and Brian Bowe. "A Multi-Level Review of Engineering Ethics Education: Toward a Socio-Technical Orientation of Engineering Education for Ethics." *Science and Engineering Ethics* 27, no. 5 (2021): 60. https://doi.org/10.1007/s11948-021-00333-6.

Martin, Vanessa. "NURI, FAŻL-ALLĀH." In *Encyclopædia Iranica*, online edition. Accessed April 30, 2024. Available at http://www.iranicaonline.org/articles/nuri-fazl-allah

Matin-asgari, Afshin. *Iranian Student Opposition to the Shah*. Costa Mesa, CA: Mazda, 2002.

Matin-Asgari, Afshin. *Both Eastern and Western: An Intellectual History of Iranian Modernity*. Cambridge: Cambridge University Press, 2018.

Matthews, Tracye A. "No One Ever Asks What a Man's Role in the Revolution Is: Gender Politics and Leadership in the Black Panther Party, 1966–1971." In *Sisters in the Struggle*, edited by Bettye Collier-Thomas and V. P. Franklin, 230–56. New York, USA: New York University Press, 2020. https://doi.org/10.18574/nyu/9780814790380.003.0017.

McKinney de Royston, Maxine, and Tesha Sengupta-Irving. "Another Step Forward: Engaging the Political in Learning." *Cognition and Instruction* 37, no. 3 (2019): 277–284. https://doi.org/10.1080/07370008.2019.1624552.

Mehan, Hugh, Irene Villanueva, Lea Hubbard, Angela Lintz, and Dina Okamoto. *Constructing School Success: The Consequences of Untracking Low Achieving Students*. New York: Cambridge University Press, 1996. https://doi.org/10.1017/CBO9781139174664.

Menashri, David. *Education and the Making of Modern Iran*. Ithaca: Cornell University Press, 1992.

Mirsepassi, Ali. *Iran's Quiet Revolution: The Downfall of the Pahlavi State*. Cambridge, United Kingdom: Cambridge University Press, 2019.

Moghissi, Haideh. *Populism and Feminism in Iran: Women's Struggle in a Male-Defined Revolutionary Movement*. New York: St. Martin's Press, 1994.

Mohammadi, Malakeh. "Pirūz bād paykār pur shawr zanān irān dar rāh āzādi, 'istiqlāl milli, va pishraft 'ijtimā'ī [Victory Upon the Passionate Struggle of Iranian Women in The Path of Freedom, National Independence, and Social Progress]." Donya 12, 1979.

Mojab, Shahrzad. "The State and University: The 'Islamic Cultural Revolution' in the Institution of Higher Education of Iran, 1980–1987." Ph.D diss., University of Illinois at Urbana-Champaign, 1991.

Mojab, Shahrzad. "State-University Power Struggle at Times of Revolution and War in Iran." *International Higher Education* 36 (2004): 11–3.

Moore, Alfred. "Architects and Engineers: Two Types of Technocrat and Their Relation to Democracy." *Critical Review (New York, N.Y.)* 32, no. 1–3 (2020): 164–181. https://doi.org/10.1080/08913811.2020.1857610.

Moradian, Manijeh. *This Flame Within: Iranian Revolutionaries in the United States*. North Carolina: Duke University Press, 2022.

Morgan, D. L., Davis, K. B., & López, N. (2020). "Engineering Political Fluency: Identifying Tensions in the Political Identity Development of Engineering Majors." *Journal of Engineering Education* 109, no. 1: 107–124.

Morales-Doyle, Daniel, and Eric "Rico" Gutstein. "Racial Capitalism and STEM Education in Chicago Public Schools." *Race, Ethnicity and Education* 22, no. 4 (2019): 525–544. https://doi.org/10.1080/13613324.2019.1592840.

Sāzmān-i Cherīkhā-yi Fadā'ī Khalq-i Īrān. Nabard Khalq, Issue. 6, 1354 [1975].

Nakhaei, Hadi. *Murur-i tawsīfī-tahlīlī taḥavulāt panjāh sāl junbish dānishjū'ī īrān: Paydāyish, gustarish, ufūl (1309–1359)*. Tehran: Pazhuhishkadeh Tarikh islam, 2016.

National Foundation for American Policy. (2022). *Analysis of U.S. and Canadian international student data*. National Foundation for American Policy. Retrieved from https://nfap.com/studies/analysis-of-u-s-and-canadian-international-student-data/

Nasir, N. S., Lee, C. D., Pea, R., & de Royston, M. M. (2020). *Handbook of the Cultural Foundations of Learning*. In Routledge eBooks (1st edition., vol. 1). Routledge. https://doi.org/10.4324/9780203774977

Nasir, Na'ilah Suad, and Victoria M. Hand. "Exploring Sociocultural Perspectives on Race, Culture, and Learning." *Review of Educational Research* 76, no. 4 (2006): 449–475. https://doi.org/10.3102/00346543076004449.

Nasr, Seyyed Hossein. "Contemporary Man, between the Rim and the Axis." *Studies in Comparative Religion* 7, no. 2 (1973): 113–126.

Nasr, Seyyed Hossein. *The Need for a Sacred Science*. Albany: State University of New York Press, 1993.

Nasr, Seyyed Hossein. "Islam and the Problem of Modern Science." *Islam & Science* 8, no. 1 (2010): 63–74.

Nasr, Seyyed Hossein. *Ḥikmat va siyāsat: Khāṭirāt duktur siyyid ḥussayn nasr*. Edited by Hossein Dehbashi. Tehran: Saziman Asnad wa Kitabkhaneh Melli Iran, 2014

Nasr, Seyyed Hossein, and Ramin Jahanbegloo. *In Search of the Sacred: A Conversation with Seyyed Hossein Nasr on His Life and Thought*. Santa Barbara, CA: Praeger, 2010.

Nasrabadi, Manijeh. "'Women Can Do Anything Men Can Do': Gender and the Affects of Solidarity in the U.S. Iranian Student Movement, 1961–1979." *Women's Studies Quarterly* 42, no. 3–4 (2014): 127–145. https://doi.org/10.1353/wsq.2014.0046.

Nasrabadi, Manijeh, and Afshin Matin-asgari. "The Iranian Student Movement and the Making of Global 1968." In *The Routledge Handbook of the Global Sixties*, 1st ed., edited by Chen Jian Martin Klimke, Masha Kirasirova, Mary Nolan, Marilyn Young, and Joanna Waley-Cohen, 443–56. United Kingdom: Routledge, 2018. https://doi.org/10.4324/9781315150918-41.

Noble, David F. *America by Design: Science, Technology, and the Rise of Corporate Capitalism*. 1st ed. New York: Knopf, 1977.

Nolan, Janice M. "The Role of International Educational Exchange Programs as a U.S. Foreign Policy Tool: A Case Study of Fulbright Program Advisers at Massachusetts Colleges in a Climate of New Nationalism." PhD diss., ProQuest Dissertations Publishing, 2018.

Nxumalo, Fikile, and Kihana Miraya Ross. "Envisioning Black Space in Environmental Education for Young Children." *Race, Ethnicity and Education* 22, no. 4 (2019): 502–524. https://doi.org/10.1080/13613324.2019.1592837.

Packer, Martin J. *The Science of Qualitative Research*. New York: Cambridge University Press, 2010.

Pahlavi, Mohammad Reza. *Ma'mūriyat barāy-i vatanam*. n.p. 1970.

Pahlavi, Mohammad Reza. *Bi sūy-i tamadun buzurg*. Frankfurt: Alborz, 2009.

Pari. "Naqsh-i dukhtarān dar fa'ālīyat-hā-yi ṣinfī- siyāsī [The Role of Females in Political and Union Activism]." *Payām-i dānishjū* 2 (n.d.): 1–15.

Paris, Django, and H. Samy Alim, eds. *Culturally Sustaining Pedagogies: Teaching and Learning for Justice in a Changing World.* New York, NY: Teachers College Press, 2017.

Partovi, Firooz. Lecture at Sharif University on the Occasion of the Fiftieth Anniversary of Its Establishment. 2016.

Pawley, Alice L. "Universalized Narratives: Patterns in How Faculty Members Define 'Engineering.'" *Journal of Engineering Education* 98, no. 4 (2009): 309–319.

Ḥezb Tūdah Īrān. Paykār: Nashriyah Ḥezb Tūdah Īrān barā-yi Dānishjūyān, vol 1, issue 1, 1971.

Ḥezb Tūdah Īrān. Paykār: Nashriyah Ḥezb Tūdah Īrān barā-yi Dānishjūyān, vol 1, issue 2, 1971.

Pensado, Jaime M. *Rebel Mexico: Student Unrest and Authoritarian Political Culture during the Long Sixties.* Stanford, CA: Stanford University Press, 2013.

Philip, Thomas M., Ayush Gupta, Andrew Elby, and Chandra Turpen. "Why Ideology Matters for Learning: A Case of Ideological Convergence in an Engineering Ethics Classroom Discussion on Drone Warfare." *The Journal of the Learning Sciences* 27, no. 2 (2018): 183–223. https://doi.org/10.1080/10508406.2017.1381964.

Pourjavadi, Nasrollah. "Ḥussayn nasr āqāzgar islāmī kardan dānishgāh-hā būd." *Kitabnamah* 20 (2013). Accessible at: http://tarikhirani.ir/fa/news/3505 /%D9%BE%D9%88%D8%B1%D8%AC%D9%88%D8%A7%D8%AF%DB%8C -%D8%AD%D8%B3%DB%8C%D9%86-%D9%86%D8%B5%D8%B1-%D8%A2%D8% BA%D8%A7%D8%B2%DA%AF%D8%B1-%D8%A7%D8%B3%D9%84%D8%A7%D9 %85%DB%8C-%DA%A9%D8%B1%D8%AF%D9%86-%D8%AF%D8%A7%D9%86% D8%B4%DA%AF%D8%A7%D9%87-%D9%87%D8%A7-%D8%A8%D9%88%D8%AF

Pouyan, Amir Parviz. *Zarurat-i mubārizah musalahānah wa radd-i ti'ūrī baqā.* n.p., 1970.

Rahnema, Ali. *Call to Arms: Iran's Marxist Revolutionaries: Formation and Evolution of the Fada'is, 1964–1976.* London: Oneworld Academic, 2021.

Rankin, Yolanda A., and Jakita O. Thomas. "The Intersectional Experiences of Black Women in Computing." In *Proceedings of the 51st ACM Technical Symposium on Computer Science Education*, 199–205. February 2020.

Redden, Elizabeth. "Report Focuses on Graduate International Enrollment." *Inside Higher Ed.* August 18, 2021. Accessible: https://www.insidehighered.com/quicktakes /2021/08/19/report-focuses-graduate-international-enrollment#:~:text=A%20 new%20report%20on%20international,percent%20in%20computer%20and%20 information

Renehan, Colm. "Peace Activism at the Massachusetts Institute of Technology from 1975 to 2001: A Case Study." Unpublished PhD diss., Boston College, 2007.

Rogoff, Barbara. *The Cultural Nature of Human Development*. Oxford: Oxford University Press, 2003.

Rūzigār sharīf. Edited by Mohammad Mirzaei. Tehran: San'ati Sharif University Press, 2016.

Ryoo, Jean J., Tiera Tanksley, Cynthia Estrada, and Jane Margolis. "Take space, make space: How students use computer science to disrupt and resist marginalization in schools." *Computer Science Education* 30, no. 3 (2020): 337–361.

Said, Edward W. *Orientalism*. New York, NY: Vintage Books, 1979.

Sayer, Derek. *The Violence of Abstraction: The Analytic Foundations of Historical Materialism*. Oxford: Basil Blackwell, 1987.

Seirafi Nejad, Mohsen. *Khāṭirāt dar sāzimān-i fadāī khalq-i īrān*. Unpublished manuscript.

Serafi Nejad, Mohsen. "Qātil panjishāhī va bīmārī kūdakī chapravī." n.p., n.d. Accessed September 1, 2023. https://iran-archive.com/sites/default/files/2022-07/moh sen-seirafinazhad-ghatle-panjeshahi.pdf.

Shahidian, Hammed. "The Iranian Left and the 'Woman Question' in the Revolution of 1978–79." *International Journal of Middle East Studies* 26, no. 2 (1994): 223–247. https://doi.org/10.1017/S0020743800060220.

Sharīf az āqāz tā kunūn bi rivāyat 'asātīd. Edited by Mohammad Mirzaei. Tehran: San'ati Sharif University Press, 2009.

Sharīf az āqāz tā kunūn bi rivāyat ru'asāy-i ān. Edited by Mohammad Mirzaei. Tehran: San'at Sharif University Press, 2006.

Shannon, Matthew K. *Losing Hearts and Minds: American-Iranian Relations and International Education during the Cold War*. Ithaca: Cornell University Press, 2017.

Showkat, Hamid, and Mehdi Khanbaba-Tehrani. *Nigāhī az darūn bi junbish chap īrān [The Iranian left: A look from within]*. Paris: Baztab, 1989.

Slaton, Amy E. *Race, Rigor, and Selectivity in U.S. Engineering: The History of an Occupational Color Line*. Cambridge, MA: Harvard University Press, 2010.

Smolansky, Oles M. "Soviet Policy in Iran and Afghanistan." *Current History* 80, no. 468 (1981): 321–339. http://www.jstor.org/stable/45315015.

Sohrabi, Naghmeh. "When Love Was Forbidden: Sex and Intimacy in Iran's Revolutionary Generation." American Academy, February 26, 2021. https://www.americanacademy.de/event/when-love-was-forbidden-sex-and-intimacy-in-irans-revolutionary-generation/.

Sohrabi, Naghmeh. "Writing Revolution as If Women Mattered." *Comparative Studies of South Asia, Africa, and the Middle East* 42, no. 2 (2022): 546–550. https://doi.org/10.1215/1089201X-9988048.

Stecopoulos, Harry, and Michael Uebel. *Race and the Subject of Masculinities*. Durham: Duke University Press, 1997.

Stevens, R., K. O'connor, L. Garrison, A. Jocuns, and D. M. Amos. "Becoming an Engineer: Toward a Three Dimensional View of Engineering Learning." *Journal of Engineering Education* 97, no. 3 (2008): 355–368.

Subramanian, Ajantha. *The Caste of Merit: Engineering Education in India*. Cambridge, MA: Harvard University Press, 2019.

Sumsion, Jennifer. "Becoming, Being and Unbecoming an Early Childhood Educator: A Phenomenological Case Study of Teacher Attrition." *Teaching and Teacher Education* 18, no. 7 (2002): 869–885. https://doi.org/10.1016/S0742-051X(02)00048-3.

Taqizadeh, Bahman. *Nigāhī az darūn bih sāzimān chirīk-hā-yi fadā'ī khalq, tārīkhchah gurūh-i munsha'ib*. Tehran: Nei, 2021.

Tan, E., A. Calabrese Barton, and A. Benavides. "Engineering for Sustainable Communities: Epistemic Tools in Support of Equitable and Consequential Middle School Engineering." *Science Education* 103, no. 4 (2019): 1011–1046.

Tavakoli-Targhi, Mohammad. *'Ayīn dānishjūyān 1323–1324, nukhustīn nashriyah dānishju'ī dānishgāh tihrān*. Toronto: Irannameh, 2016.

The Tech. "Selling MIT, Bombs for the Shah." *The Tech*, March 7, 1975, 4.

The Tech. "Iranian Program Attacked." *The Tech*, March 18, 1975, 3.

The Tech. "Nuc-Eng Sit-In Held to Protest Iran Deal." *The Tech*, April 29, 1975, 1.

The Tech. "Kindleberger Named the Head Iran Panel." *The Tech*, May 9, 1975, 1.

The Tech. "Iran Group Proposal Killed." *The Tech*, November 21, 1975, 1.

The Tech. "Iranians Disrupt Colby Talk." *The Tech*, April 14, 1978.

Tetlock, Philip E., Randall S. Peterson, Charles McGuire, Shi-jie Chang, and Peter Feld. "Assessing Political Group Dynamics: A Test of the Groupthink Model." *Journal of Personality and Social Psychology* 63, no. 3 (1992): 403–425. https://doi.org/10.1037/0022-3514.63.3.403.

Toliver, Stephanie R. "Can I Get a Witness? Speculative Fiction as Testimony and Counterstory." *Journal of Literacy Research* 52, no. 4 (2020): 507–529. https://doi.org/10.117 7/1086296X20966362.

Vahabzadeh, Peyman. *A Guerrilla Odyssey: Modernization, Secularism, Democracy, and Fadai Period of National Liberation in Iran, 1971–1979*. Syracuse, NY: Syracuse University Press, 2010.

Vahabzadeh, Peyman. "Secularism and the Iranian Militant Left: Political Misconception or Cultural Issues?" *Comparative Studies of South Asia, Africa, and the Middle East* 31, no. 1 (2011): 85–93. https://doi.org/10.1215/1089201X-2010-055.

Vahabzadeh, Peyman. "FADĀʾIĀN-E ḴALQ." In *Encyclopædia Iranica*, online edition, 2015. Available at http://www.iranicaonline.org/articles/fadaian-e-khalq (accessed April 29, 2024).

Vakil, Sepehr. "Ethics, Identity, and Political Vision: Toward a Justice-Centered Approach to Equity in Computer Science Education." *Harvard Educational Review* 88, no. 1 (2018): 26–52.

Vakil, Sepehr. "'I've Always Been Scared That Someday I'm Going to Sell Out': Exploring the Relationship between Political Identity and Learning in Computer Science Education." *Cognition and Instruction* 38, no. 2 (2020): 87–115. https://doi.org/10.1080/07370008.2020.1730374.

Vakil, Sepehr, and Elham Beheshti. "Theorizing the Politics of Identity in Engineering: Reflections From the University of Tehran, Iran." In 14th International Conference of the Learning Sciences: The Interdisciplinarity of the Learning Sciences, ICLS 2020 (pp. 729–732). International Society of the Learning Sciences (ISLS).

Van Manen, Max. *Researching Lived Experience: Human Science for an Action Sensitive Pedagogy*. London, Ont: Althouse Press, 1990.

Vossoughi, Shirin. "Elsewhere Worlds, Poetics and the Science of Human Learning." Jan Hawkins Award Lecture, 2021.

Vossoughi, Shirin, Ava Jackson, Suzanne Chen, Wendy Roldan, and Meg Escudé. "Embodied Pathways and Ethical Trails: Studying Learning in and through Relational Histories." *The Journal of the Learning Sciences* 29, no. 2 (2020): 183–223. https://doi.org/10.1080/10508406.2019.1693380.

Vossoughi, Shirin, and Kris D. Gutiérrez. "Critical Pedagogy and Sociocultural Theory." In *Power and Privilege in the Learning Sciences*, edited by Indigo Esmonde and Angela N. Booker, 157–79. New York: Routledge, 2017.

Vossoughi, Shirin, and Sepehr Vakil. "Toward What Ends?: A Critical Analysis of Militarism, Equity, and STEM Education." In *Education at War*, 1st ed., edited by Arshad Imtiaz Ali and Tracy Lachica Buenavista, 117–40. New York, USA: Fordham University Press, 2018. https://doi.org/10.1515/9780823279111-007

Vygotsky, Lev Semenovich. *The Cambridge Companion to Vygotsky*. Edited by Harry Daniels, Michael Cole, and James V. Wertsch. Boston, MA: Cambridge University Press, 2007.

Warren, Beth, Shirin Vossoughi, Ann S. Rosebery, Megan Bang, and Edd V. Taylor. "Multiple Ways of Knowing: Re-Imagining Disciplinary Learning." In *Handbook of the Cultural Foundations of Learning*, edited by Na'ilah Suad Nasir, Carol D. Lee, Roy D. Pea, and Maxine McKinney de Royston, 277–94. Routledge, 2020. https://doi.org/10.4324/9780203774977-19.

Wisnioski, Matthew H. *Engineers for Change: Competing Visions of Technology in 1960s America*. Cambridge, MA: MIT Press, 2012.

Wisnioski, Matthew H. "Why MIT Institutionalized the Avant-Garde: Negotiating Aesthetic Virtue in the Postwar Defense Institute." *Configurations* 21, no. 1 (2013): 85–116.

Wortham, S. E. F. *Learning Identity: the Joint Emergence of Social Identification and Academic Learning.* Cambridge, MA: Cambridge University Press, 2006.

Yādigārān (kitāb 'aks sharīf), Edited by Mohammad Mirzaei. Tehran: San'ati Sharif University Press, 2005.

Yates, Miranda, and James Youniss. "Community Service and Political Identity Development in Adolescence." *Journal of Social Issues* 54, no. 3 (1998): 495–512. https://doi.org/10.1111/j.1540-4560.1998.tb01232.x.

Yazdi, Ebrahim. *Junbish-i dānishjū'ī dar dū dahah az 1320 tā 1340.* Tehran: Qalam, 2004.

Zahidi, Saadia. *Fifty Million Rising: The New Generation of Working Women Transforming the Muslim World.* 1st ed. New York: Nation Books, 2018.

Zarghamee, Mehdi. "Mojtahedi and the Founding of the Arya-Mehr University of Technology." *Iranian Studies* 44, no. 5 (2011): 767–775. http://www.jstor.org/stable/23033299.

Zavala, Miguel. "Design, Participation, and Social Change: What Design in Grassroots Spaces Can Teach Learning Scientists." *Cognition and Instruction* 34, no. 3 (2016): 236–249. https://doi.org/10.1080/07370008.2016.1169818.

Zirinsky, Michael P. "A Panacea for the Ills of the Country: American Presbyterian Education in Inter-War Iran." *Iranian Studies* 26, no. 1–2 (1993): 119–137. https://doi.org/10.1080/00210869308701789.

Ziyazarifi, Abulhassan. *Sāzimān dānishjūyān dānishgāh tihrān.* Tehran: Shirazeh, 2016.

ARCHIVES

Aryamehr University of Technology Archives, cited as Ganjīnah Historical Document Center.

Harvard Oral History Project, cited as HOHP.

Massachusetts Institute of Technology Archives, cited as MIT.

National Library and Archives of Iran, cited as NLAI.

Parliamentary Library of Iran, cited as PLI.

INDEX

Note: Photos are indicated by italicized page numbers.